Programming the BBC micro:bit

Getting Started with MicroPython

Simon Monk

New York   Chicago   San Francisco
Athens   London   Madrid
Mexico City   Milan   New Delhi
Singapore   Sydney   Toronto

Library of Congress Control Number: 2017954427

Programming the BBC micro:bit: Getting Started with MicroPython

1 2 3 4 5 6 7 8 9   LCR   22 21 20 19 18 17

ISBN   978-1-260-11758-5
MHID   1-260-11758-8

Sponsoring Editor
Michael McCabe

Editorial Supervisor
Stephen M. Smith

Production Supervisor
Lynn M. Messina

Project Manager
Patricia Wallenburg,
TypeWriting

Copy Editor
James Madru

Proofreader
Claire Splan

Indexer
Claire Splan

Art Director, Cover
Jeff Weeks

Composition
TypeWriting

To my mother Anne Kemp,
whose kindness and positive attitude to life
are an example for all who know her.

About the Author

Simon Monk (Preston, UK) has a bachelor's degree in cybernetics and computer science and a Ph.D. in software engineering. He has been an active electronics hobbyist since his early teens and since 2012 has divided his work life between writing books and designing products for the business he started with his wife (http://monkmakes.com), which manufactures hobby electronics kits and boards.

You can find out more about Simon's books at http://simonmonk.org. You can also follow him on Twitter, where he is @simonmonk2.

CONTENTS

ACKNOWLEDGMENTS

I'd like to thank my editor, Mike McCabe, and everyone at TAB/McGraw-Hill Education for being such a great publisher to work for. Special thanks are due to Patty Wallenburg for her wonderful organizational skills, keen eye, and ability to make my books look as good as they possibly can.

Many thanks to David Whale for taking the time to complete a detailed and helpful technical review of the book and to all at the micro:bit foundation who have been friendly and helpful during the writing of this book.

Once again, thanks to Linda for believing in me and giving me the support and encouragement to make a success of writing.

Simon Monk

1

Introduction

The micro:bit (Figure 1-1) is a small uncased circuit board with a display made up of 25 LEDs, a couple of buttons, and some sensors. Crucially, it has a micro-USB socket that allows you to connect it to your computer both to power it and to send programs to it. The micro:bit can also be connected to a battery pack so that it can be used without your computer.

The micro:bit was designed for use in education, but this useful little device has also endeared itself to electronics hobbyists and makers around the world.

Plug Me In!

To use your micro:bit, you will need a USB to micro-USB lead to connect it to your computer. You may have bought one at the same time as you bought your micro:bit. But if not, don't worry because this lead is probably the most common USB lead in existence. The micro-USB connector on the end that plugs into the micro:bit is the same plug as is used on most non-Apple cellphones and countless other electronics devices. You will occasionally come across USB leads that are "charge only." This means that they do not have the necessary wires inside to provide a data connection for the micro:bit. So, if you have problems when it comes to connecting to your micro:bit, try changing the lead for one that claims to be a "data" lead.

Figure 1-2 shows a micro:bit attached to a laptop. This laptop happens to be an Apple, but the micro:bit can be used with a Windows, Linux, Apple, or even Raspberry Pi computer.

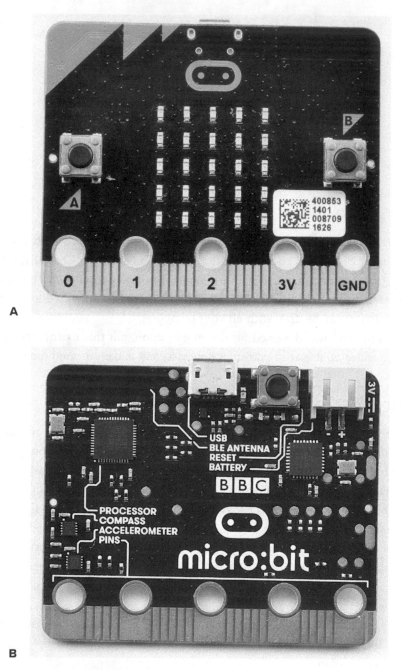

Figure 1-1 *The micro:bit: (A) front; (B) back.*

Figure 1-2   *Connecting your micro:bit to your computer.*

Plug your new micro:bit in, and a little animation will start that displays a welcome "Hello" message and then goes on to point out the two A and B buttons built into the board. Take some time to familiarize yourself with the board, and follow the prompts on the scrolling display.

If you want to start again, just press the Reset button (see Figure 1-1*B*) just to the right of the the USB connector.

History

The micro:bit is more correctly called the *BBC micro:bit*. The BBC (British Broadcasting Corporation) is the United Kingdom's largest public-service broadcaster. The BBC micro:bit project was designed to provide an easy-to-use platform to teach children how to code while realizing that coding can also be used to control electronics and not just make things happen on screens.

As part of this initiative, in 2016, around a million micro:bits were given out free of charge to every school child in UK school year 7 (11- or 12-year-olds) in the UK public school system. Since then, the running of the micro:bit project has passed from the BBC to the Microbit Educational Foundation. This not-for-profit organization is now spreading the use of the micro:bit to educational communities around the world and also making this handy little device available to electronics hobbyists and makers.

What Can It Do?

When you plug in your micro:bit, it will give you a quick rundown of its own features. Let's go through these features in a little more detail.

- **LED display.** This display is made up of 25 LEDs arranged in a 5 × 5 grid. You can control the brightness of any of these LEDs separately and display text messages on the screen that scroll from right to left.

- **Push buttons A and B.** You can write programs that will perform certain actions (perhaps display a message) when one of these buttons is pressed.

- **Touch pins.** The connectors marked 0, 1, and 2 (at the bottom of Figure 1-1*A*) can be used as touch switches, so by touching them you can trigger actions in your programs. You can also use alligator clips to attach these connectors to objects so that your programs can detect you touching those objects. Figure 1-3 shows a micro:bit attached to a banana!

- **Accelerometer.** An accelerometer measures acceleration—that is, the rate of change of speed. It does this by measuring the forces acting on a little weight built into a chip. There is such a chip on the micro:bit, and it allows you to do such things as detect the orientation of the micro:bit (like a Wii controller) or whether it is being shaken or just moved slightly, opening up all sorts of project opportunities.

- **Compass.** The micro:bit also has a built-in electronic compass that lets you detect the direction in which it is facing or detect whether you place a magnet near it.

- **Radio.** The micro:bit has a radio transmitter/receiver that can be used to send messages from one micro:bit to another.

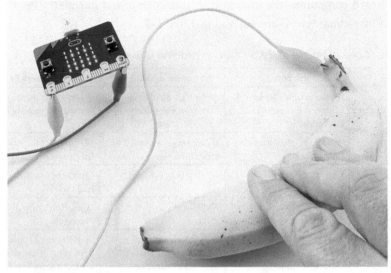

Figure 1-3 *micro:bit and fruit.*

micro:bit versus Raspberry Pi and Arduino

Keeping up with new boards such as the micro:bit can be a daunting task. It's not always immediately obvious what the differences are and which you should use. Figure 1-4 shows perhaps the three most popular microcontroller and

Figure 1-4 *(From left to right) micro:bit, Arduino Uno, and Raspberry Pi 3.*

single-board computers: the micro:bit, Arduino Uno, and Raspberry Pi. The features of these boards are summarized in Table 1-1.

Table 1-1 *Comparing the micro:bit with an Arduino Uno and Raspberry Pi*

	micro:bit	Arduino Uno	Raspberry Pi 3
General description	A microcontroller board designed for use in education	A microcontroller board for control applications and education	A low-cost single-board computer running Linux
Memory	256 kB for programs, 16 kB for data	16 kB for programs, 1 kB for data	SD card (typically 8 GB) for programs and files
Processor	Cortex M0 (16 MHz)	ATmega328 (16 MHz)	Quad-core Cortex A53 (1.2 GHz)
Built-in peripherals	Accelerometer, compass, LED display, radio	None	WiFi, Bluetooth, HDMI video out
Input/output	19 general-purpose 3.3V pins, some shared with other functions, including 6 analog inputs	20 general-purpose 5V pins, some shared with other functions, including 6 analog inputs	26 general-purpose 3.3V pins, some shared with other functions; no analog inputs
Guide price	$15	$25	$40

As you can see from this table, the micro:bit actually represents a pretty good value compared with the Arduino Uno. Where the Arduino scores is in its connectors, which allow wires to be plugged in directly, as well as a vast range of plug-in "shield" circuit boards that sit on top of the Arduino, providing extra features.

The Raspberry Pi 3 is really a very different beast, actually being a fully fledged computer to which you can attach a keyboard, mouse, and monitor. However, unlike most computers, the Raspberry Pi has general-purpose "pins" to which you can attach external electronics just like micro:bit and the Arduino.

If you look around your home, you are quite likely to find a range of devices that contain the equivalents of a micro:bit, Arduino, or Raspberry Pi. For example, your TV remote control contains a microcontroller (like the "brain" of a micro:bit or Arduino) that has pins to which the remote control's keys are connected and an infrared LED that will send out a series of pulses that transmit a code to the receiver on the TV.

A smartphone or media center will have something very similar to the processing part of a Raspberry Pi at its core. This will be a much more powerful device than the micro:bit or Arduino because it will need to be capable of generating video signals and also running an operating system that can do more than one thing at a time.

So What Is Programming?

As we have already established, many of the electronics goodies that you buy contain a microcontroller like the micro:bit's microcontroller. This is a tiny computer designed to monitor buttons and switches as *inputs* and control *outputs*. You will find microcontrollers in car radios, kitchen timers, TV remotes, and even disposable greetings cards that play a tune when you open them. Pretty much anything that has push buttons and some kind of display will also have a microcontroller in it.

Without programming, a microcontroller doesn't do anything. It will just sit there idly until someone uploads a program to it telling it what to do. Telling a microcontroller what to do involves writing a computer program (called *programming* or *coding*) and then installing the program on the microcontroller.

If you look at your TV remote, it's unlikely to have a USB port through which to program it. In cases such as this, where a large number of devices are made, special programming hardware is used to program the device while it is being manufactured. Once programmed, the device probably will never have another program installed on it. This is quite different from the micro:bit, which has a built-in programming interface that allows you to install a new program hundreds of thousands of times if you want. Figure 1-5 shows how the whole programming process works on a micro:bit when using MicroPython.

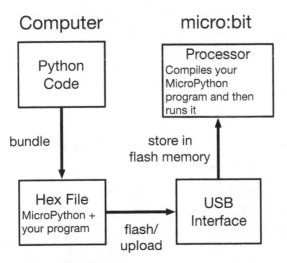

Figure 1-5 *Programming a micro:bit.*

The process starts with your program written in the Python programming language using an editor program running on your computer. When you are ready to transfer the program to the micro:bit, the editor program first converts the text version of the program along with MicroPython itself into a file in a format called *hex* (short for *hexadecimal*). This file is then transferred to the micro:bit (*uploaded* or *flashed*) simply by copying the hex file from your computer's disk drive to the micro:bit's virtual USB device. This is made possible by a neat trick of the micro:bit's to look like a data storage device to your computer. The uploaded file is then processed by the micro:bit into a binary form that it can run (called *compiling*). Once the process is complete, the program will run automatically.

Why MicroPython?

There are many computer languages with which you can program your micro:bit. The two most popular and the two being promoted by the Microbit Educational Foundation are MicroPython and JavaScript Blocks. JavaScript Blocks uses a graphical-type interface to plug together code *blocks* to put your program together, whereas MicroPython is a more conventional free text language that is completely text based. Chapter 12 provides an introduction to JavaScript Blocks.

Both languages have their pros and cons, but this book concentrates on MicroPython, which is an implementation of Python created by Damien George specifically for use on low-power microcontrollers such as the micro:bit. Python is a popular language in educational settings because it is considered to be one of the easiest programming languages to learn that is also well used in the software industry.

Summary

In this chapter, you have plugged your micro:bit into your computer and learned a little about this small device and what it can be used for. In Chapter 2, you will start to program your micro:bit.

2

Getting Started

In this chapter, you will learn how to run your first program on your micro:bit and also come to grips with the two most popular editors used to write and upload code onto the device.

MicroPython Editors

When it comes to programming the micro:bit using MicroPython, there are two main options for the editor with which to type your programs. The first is browser based. It doesn't really have a formal name, so let's just call it the *online editor*. Because it works in your browser, there is nothing to install. To use the MicroPython online editor, all you need to do is navigate to http://python.microbit.org/editor .html, and the Editor window will appear (see Figure 2-1).

There are some disadvantages to this option. For one, it is all too easy to accidentally lose the page and your code before you get around to saving it. The process of uploading the program to your micro:bit is also a bit fiddly. You also have to have an Internet connection to be able to use the editor.

The second option is an editor called *Mu* that is a normal application that you run on your computer. Mu also has a neat feature that allows you to try out individual lines of code directly on the micro:bit without having to upload an entire program.

You will begin by using the online editor and then move on to Mu. It is then up to you as to which you prefer to use for the rest of this book. The code will work equally well with both editors, but some of the coding experiments do require the use of Mu if you want to follow along.

Figure 2-1   *The MicroPython browser editor.*

Online Editor

Visit http://python.microbit.org/editor.html in your browser (you may want to bookmark this page), and you will see something like Figure 2-1. You can see that there is already a short program in the editor that will display the message "Hello, World!" on the LED display.

Across the top of the online editor is a row of toolbar buttons. The important ones are Download and Save. When you click on Download, the online editor will generate a *hex* file that is ready to flash onto your micro:bit and also download it into your computer's Downloads area. Where this Downloads area is depends on your browser and operating system. Try this out now with the code already in the online editor by clicking on Download (the leftmost button). The download won't take long because the file is small. You now have to transfer this program from your computer to your micro:bit.

When you connect a micro:bit to your computer, it "mounts" the micro:bit as if it were a USB flash drive. This makes copying the hex file from your computer to the micro:bit simply a matter of dragging the file from your computer to the micro:bit. The exact way to do this depends on your computer's operating system.

Installing micro:bit Programs Using Windows

When you click on Download on a Windows computer using the Microsoft Edge browser, you will see the message "Microbit.hex finished downloading" along with two options, Open and View Downloads (Figure 2-2).

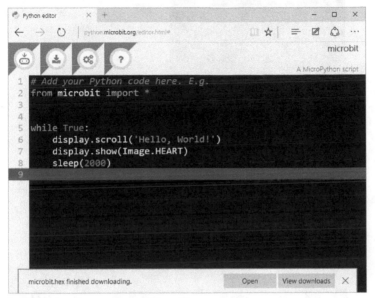

Figure 2-2 *Downloading a micro:bit hex file.*

Click on the option View Downloads, and a panel will open on the window (Figure 2-3) showing the file downloads made by the browser, including the one you just downloaded (microbit.hex).

You cannot copy this downloaded file from this window onto the micro:bit. Instead, you need a File Explorer window, which can be opened by clicking on the Open Folder option. This opens the File Explorer shown in Figure 2-4, from which you can just drag the file microbit.hex onto the micro:bit drive shown on the left of the window.

When you do this, the file copying dialog of Figure 2-5 will appear. A yellow LED on the micro:bit will flicker during the flashing. After a few seconds, the upload will be complete, the micro:bit will reset, and your micro:bit should start displaying the scrolling message.

Figure 2-3 *Viewing the file downloads.*

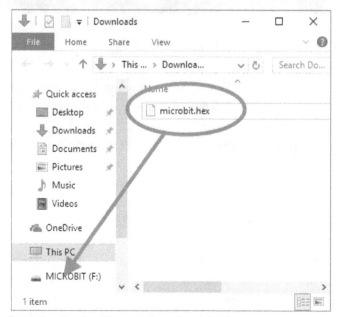

Figure 2-4 *Dragging the hex file onto the micro:bit drive.*

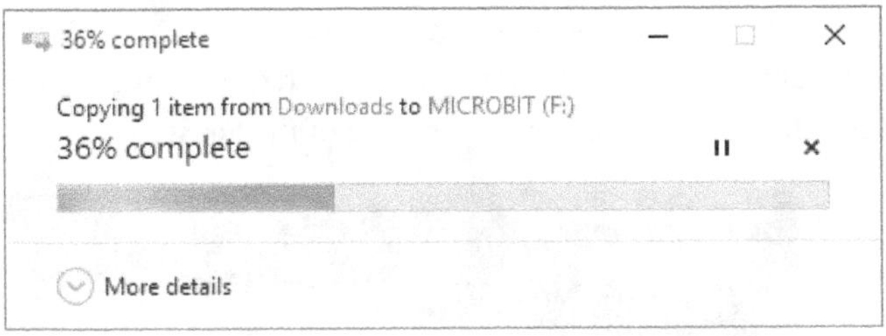

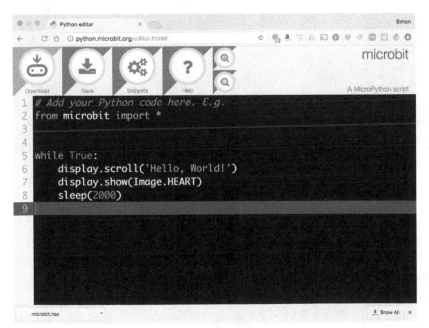

Figure 2-5 *File upload progress.*

Congratulations, you have uploaded your first program to your micro:bit! Having taken the long route to opening the File Explorer to copy hex files onto the micro:bit, you can just leave the File Explorer of Figure 2-4 open, ready to copy over another hex file.

Installing micro:bit Programs Using a Mac

Figure 2-6 shows the "Hello World!" program in Google Chrome on a Mac after clicking the Download button. You can see the downloaded file (microbit.hex) at the bottom of the window.

Figure 2-6 *Downloading a micro:bit hex file on a Mac.*

Click on the Menu button next to the download (Figure 2-7), and select the option Show in Finder. This will open the Finder with the hex file selected and ready to drag onto the micro:bit in the Devices section of the Finder window (Figure 2-8).

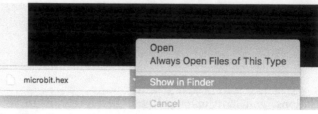

Figure 2-7 *The Download options menu.*

Figure 2-8 *Dragging the hex file onto the micro:bit drive (on a Mac).*

Installing micro:bit Programs Using Linux

Linux computers, including the Raspberry Pi, can be used with the micro:bit. Figure 2-9 shows a micro:bit connected to a Raspberry Pi.

The process is the same as for Windows and Mac. You first download the hex file onto your computer by clicking on the Download button. You then copy the downloaded hex file onto the micro:bit. The location of the downloaded file depends on which browser you are using, but for Chromium on the Raspberry Pi (Figure 2-10) and most browsers, the Downloads folder is in your home directory.

Figure 2-9 *Using micro:bit with a Raspberry Pi.*

Figure 2-10 *The online editor on a Raspberry Pi.*

When you plug your micro:bit into your Raspberry Pi or other Linux computer, you may be prompted with a dialog like the one shown in Figure 2-11 that offers to treat the micro:bit as a removable storage device and open it in the File Manager so that you can copy hex files onto it (Figure 2-12).

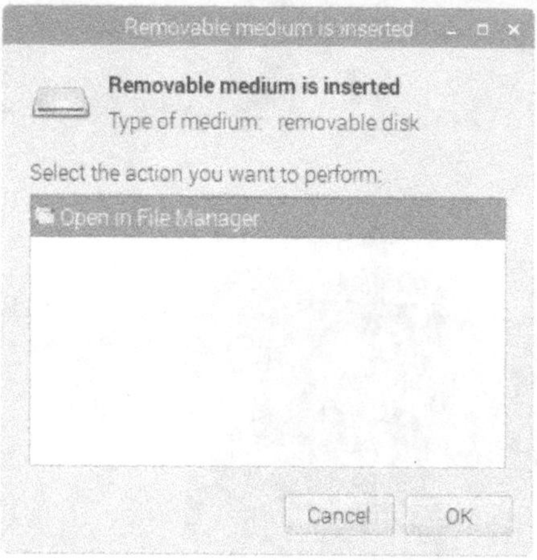

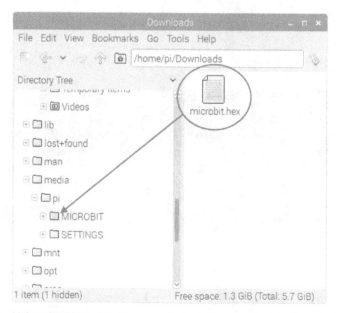

Figure 2-11 *The micro:bit mounting as removable media.*

Figure 2-12 *Using the Linux File Manager to install a program on the micro:bit.*

Saving and Loading Programs

You will find the Save button in the button area at the top of the editor window. You might expect this to offer you a file dialog to let you save the file in a particular location. However, when you click on it, what it actually does is simply download the Python script shown in the editor into your computer's Downloads area. You can then move the file somewhere else if you want to, probably after renaming it.

Loading files into the editor is simply a matter of dragging a file onto the online editor's main editing area. But be careful because this will replace any text already in the editor.

The Mu Editor

The Mu editor (Figure 2-13) makes loading/saving and deploying a lot easier, but it does require you to download the application onto your computer, and if you are using Windows, you can install a USB driver if you want to make use of Mu's advanced feature of allowing an interactive interface to the micro:bit called the *Read Eval Print Loop* (REPL).

Figure 2-13   *The Mu editor.*

Installing Mu on Windows

Click on the Download for Windows button at https://codewith.mu (Figure 2-14), which will download the Mu application itself. Move this to your Desktop folder by using the File Explorer to drag it from your Downloads directory (Figure 2-15).

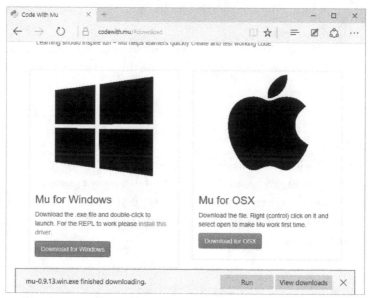

Figure 2-14 *Downloading Mu.*

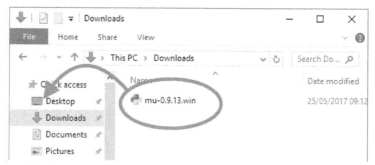

Figure 2-15 *Moving Mu to the desktop.*

To run Mu, double-click on the Mu icon on your desktop. Near the Download for Windows button on the Mu website's download page, you will also find a link called "Install this Driver." This is not essential, but it is well worth doing because it is needed if you want to experiment with the micro:bit's REPL command-line interface. This will take you to a webpage hosted by ARMmbed for instructions on downloading and installing the driver. Follow the instructions there. Note that the installer program will not install the driver without your micro:bit being connected to your computer. Note also that you will need admin privileges on your computer to install the driver.

Installing Mu on a Mac

The Mac is the easiest platform on which to install Mu. Click on the Download for OSX button at https://codewith.Mu, which will download a Zip archive file. Unzip the file by double-clicking on it, and then drag the resulting application into your Applications folder. OSX includes the USB driver used by the micro:bit, so there is no need to install anything for the REPL to work.

Installing Mu on Linux

To install Mu on your Linux computer, including the Raspberry Pi, open a terminal session, and run the command:

```
$ sudo apt-get install mu
```

Once installed, you will find a shortcut to Mu in the Programing submenu of your window manager.

Using Mu

Mu (see Figure 2-13) has a lot more buttons on its toolbar than the online editor but otherwise looks fairly similar. The main feature is a big area in the middle of the window for you to write your code. Let's try the "Hello World!" example in Mu.

When you first start Mu, it will look like Figure 2-13. Mu is editing an unsaved program with the following content:

```
from microbit import *


# Write your code here :-)
```

The first line is an `import` command that tells Mu that we need to use the "micro:bit" module that contains the micro:bit-specific code for controlling the LEDs and other hardware. The second line, which starts with a #, is a comment line.

Comments are not actually program code; they are like notes that you the programmer can leave in the code to either explain something tricky going on in the program or (as is the case here) an instruction for someone else reading the code to do something.

Replace the comment line with the following code:

```
while True:
  display.scroll('Hello, World!')
  display.show(Image.HEART)
  sleep(2000)
```

You will find that after the first line, Mu will automatically indent the next line when you press ENTER. Any line that ends with a colon (:) will indicate to Mu that the line that follows it should be indented.

You can also indent lines of code manually by pressing the TAB key before you type the first character of the line. This inserts four spaces. MicroPython is very fussy about consistency in indentation and will give error messages if it's not correct. When you have finished, the Mu window should look like Figure 2-16.

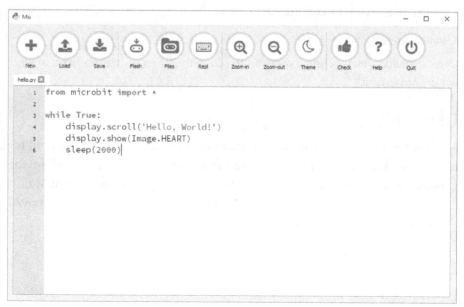

Figure 2-16   *"Hello World!" in Mu.*

Mu won't stop you from downloading faulty programs onto your micro:bit, and Mu has a Check button that is intended to check your code before uploading. At the time of writing, this checking is a little overly zealous, so it will tell you such things as "Blank line contains whitespace," which just means that there is a blank line with a few space characters on it. This means that the advice offered by the Check feature is somewhat verbose and makes it hard to pick out the genuine errors. Try it, because Mu is a young product that is improving all the time.

If your program does have errors when it is flashed onto your micro:bit, you will see the message scroll across the micro:bit's display. The most useful piece of information is often the line number of the error. But remember that an error on one line is quite often caused by an error on the preceding line.

If you open the REPL (see the next section) before you flash the program, any error messages will appear there, where they are a lot easier to read than on the scrolling display.

Now is a good time to save your program, so click on the Save button, and choose a file name (perhaps `hello.py`). The file extension `.py` indicates that the file is a Python program. By default, Mu will save your programs in a folder called `mu_code`, but you can save the programs anywhere.

You can now transfer the program onto the micro:bit by clicking on the Flash button. If your micro:bit is connected to your computer, this will install the program automatically without you having to copy the file yourself. If you forget to plug your micro:bit in, before clicking Flash, Mu will open a file dialog allowing you to save the hex file for manual installation.

The REPL

REPL gives you a command line that actually runs on your micro:bit. It can be handy for trying out MicroPython commands on the micro:bit without having to go to the trouble of writing an entire program. To try it out, plug in your micro:bit, and then click on the Repl button in the toolbar. The result is shown in Figure 2-17.

If you don't see something like Figure 2-17, then try pressing CTRL-C on your keyboard. You are now communicating with the micro:bit, which is awaiting your command. So let's have it do a bit of math for us. Type the following text into the REPL:

```
2+2
```

and then hit the ENTER key. You can see the result in Figure 2-18. Your micro:bit has added 2 and 2 together and is showing the result of 4.

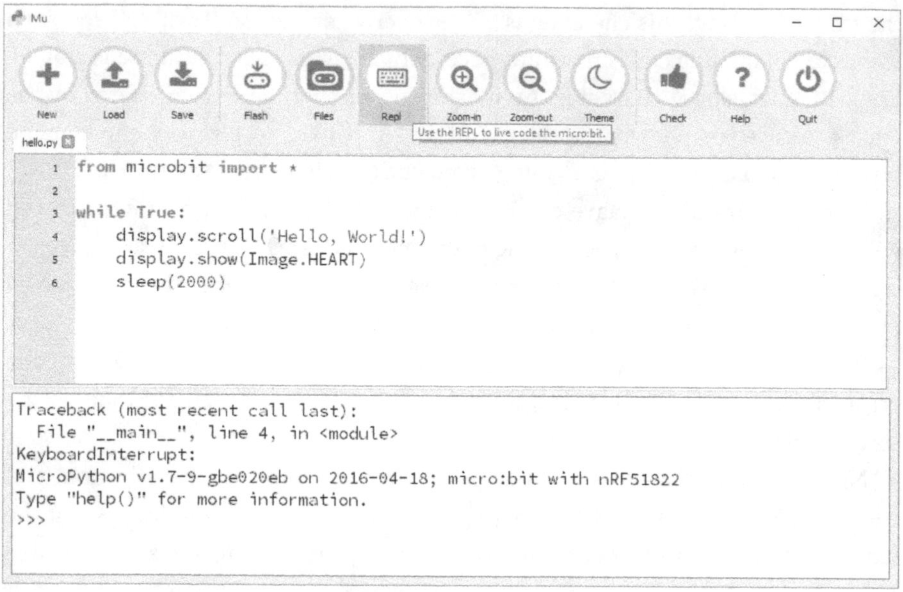

Figure 2-17 *The REPL.*

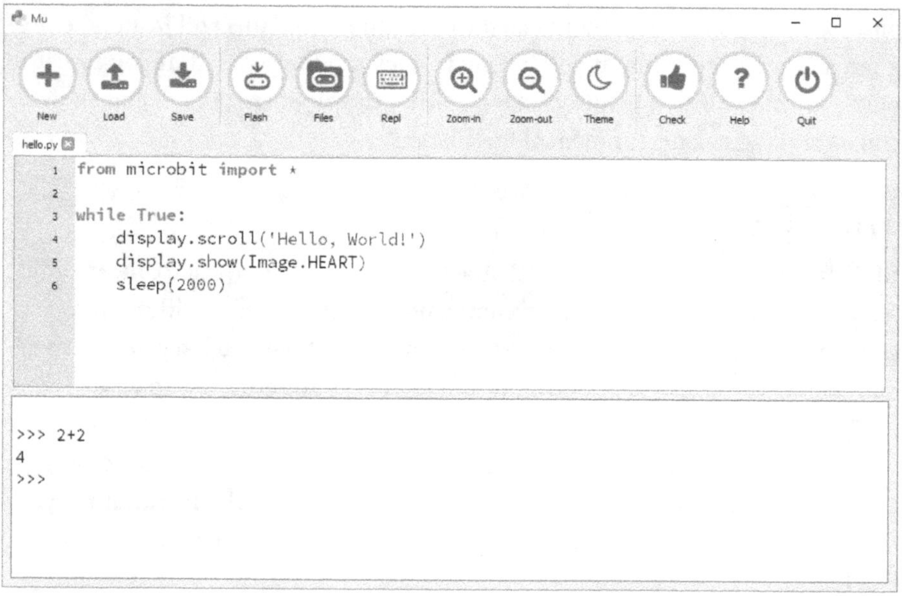

Figure 2-18 *Sums in the REPL.*

Next, let's try using the display to scroll a message from the REPL. To do this, you need to type the following two lines of code:

```
from microbit import *
display.scroll('Programming the micro:bit')
```

Be careful not to forget the single-quote marks around the text to be displayed. The message should appear on your micro:bit's display as soon as you press ENTER after the second line.

You will find the first line of code at the top of all micro:bit MicroPython programs, and it "imports" all the micro:bit-specific code for things such as the display and buttons into the MicroPython environment, ready for you to use.

Downloading This Book's Programs

Although it can be a good thing to actually copy in simple code examples by hand, as the programs in this book get longer, you definitely don't want to be typing them in by hand. So all the programs used in this book can be downloaded from the book's Github repository.

Now is a good time to download all the code examples so that they are there when you need them. To do this, go to https://github.com/simonmonk/prog_mb with your browser. Click on the green Clone or Download button, and select the option Download Zip (Figure 2-19). Unzip the file, which will create a folder called prog_mb. This will contain all the programs. To load one of these programs into the online editor, just drag the file onto the Editor area.

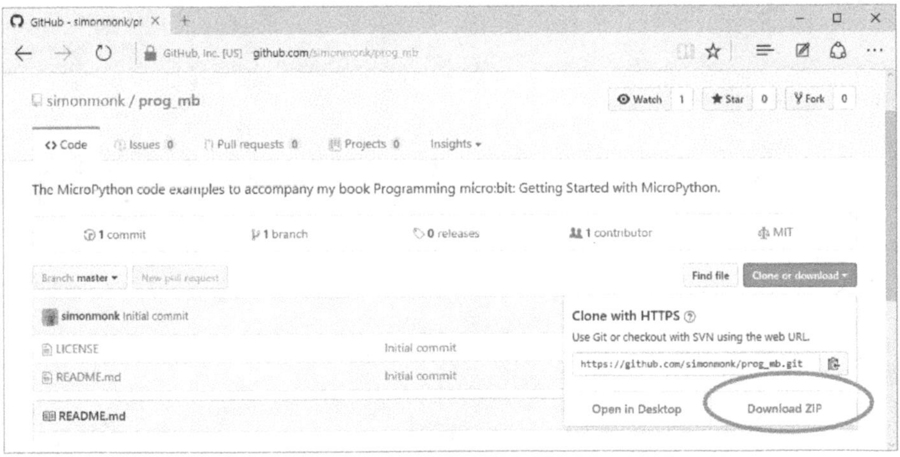

Figure 2-19 *Downloading the example programs.*

If you are using Mu, then it is worth moving the files to the special directory where Mu expects to find your programs. This is in a directory called mu_code in your home directory. For example, in Windows 10, this is in C:\Users\ YourName\mu_code. So move the downloaded and unzipped files from prog_mb into mu_code.

To load a file into Mu, click on the Load button, and navigate to the file you want to open.

Summary

Now that your computer is set up for the micro:bit and you know how to connect a micro:bit and program it, we can turn our attention to MicroPython. In Chapter 3 you will start to explore this programming language.

3

MicroPython Basics

In this chapter, you will learn a bit about programming in MicroPython and get your micro:bit running a few programs. MicroPython is one of many implementations of the Python language, and it is designed specifically for low-power microcontroller boards such as the micro:bit. All the general information in this book about MicroPython is equally applicable to other implementations of Python. The real differences arise when it comes to using the micro:bit hardware.

Numbers

Remember the Read Eval Print Loop (REPL) from Chapter 2? Well let's start by carrying out a few experiments using the REPL. Start Mu, connect your micro:bit to your computer, and click on the REPL button. Start by repeating the REPL experiment of Chapter 2 and type 2 + 2 after the >>> prompt to see the result shown in Figure 3-1.

Note that when you use the REPL like this, it doesn't matter what program you have loaded onto your micro:bit. It will be halted while you use the REPL and only restart if you reset the micro:bit by pressing the RESET button, unplugging it and plugging it back in, or uploading a new program. The Traceback message that you see before the >>> prompt is a result of the halting of whatever program was running.

There are two important types of numbers in Python: *integers* (or *ints* for short), which are whole numbers such as 1, 2, 3, and so on, and *floats* (*floating points*), which have a decimal place such as 1.1, 2.5, 10.5, and so on. In many

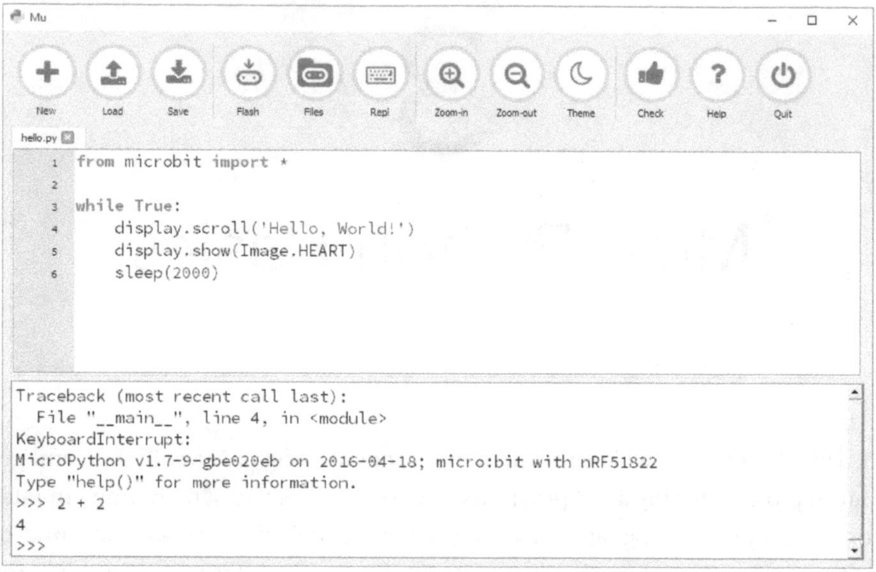

Figure 3-1 *2 + 2 in the REPL.*

situations, you can use them interchangeably. Now try typing the following after the REPL prompt:

```
10 + 5.5
```

As you can see, the result is 15.5, as you would expect. Try out a few sums for yourself. If you want to do multiplication, use * and for division use /.

Doing some simple sums on your micro:bit is all very well, but you could do that on a pocket calculator. Note that from now on, the things you need to type will be proceeded by >>>, and the micro:bit's response will start on a new line.

Enter the following into the REPL:

```
>>> from random import randint
>>> randint(1, 6)
5
```

You are likely to get a different answer than 5. Repeat the randint(1, 6) line a few times by pressing the up arrow on your keyboard, and you will see a succession of random numbers between 1 and 6. You have made a die of sorts!

The first line you typed imports the randint function from the module random. Modules are used to contain Python code that you may not need all the time. This helps to keep the size of the code small enough to run on a micro:bit.

Variables

Try typing the following line into the REPL:

```
>>> x = 10
```

You can put spaces either side of the equals sign (=) or not—it's a matter of personal preference. I think it looks neater with spaces, so that's the standard I will be sticking to in this book.

This line of code uses a *variable* called x. Now that the variable x has been given the value 10, try just typing x in the REPL:

```
>>> x
10
```

Python is telling us that it remembers that the value of x is 10. You don't have to use single-letter names for variables; you can use any word that starts with a letter, but variables can include numbers and the underscore character (_). Sometimes you need a variable name that's made up of more than one human language word. For example, you might want to use the name my number rather than x. You can't use spaces in variable names, so you use the underscore character to join the words like this: my_number.

By convention, variables usually start with a lowercase letter. If you see variables that start with an uppercase letter, it usually means that they are what are called *constants*. That is, they are variables whose value is not expected to change during the running of the program but that you might want to change before you flash your program onto your micro:bit. For example, in Chapter 11, the program ch11_messenger.py uses a constant called MY_ID that must be set to a different number for each micro:bit onto which it is flashed.

Returning to our experiments with REPL, now that Python knows about x, you can use it in sums. Try out the following examples:

```
>>> x + 10
20
>>> x = x + 1
>>> x
11
```

In this last command (x = x + 1), we have added 1 to x (making 11) and then assigned the result to x, so x is now 11. Increasing a variable by a certain amount

is such a common operation that there is a shorthand way of doing it. Using +=
combines the addition and the assignment into one operation. Try the following,
and x will now be 21.

```
>>> x += 10
>>> x
21
```

You can also use parentheses to group together parts of an arithmetic expres-
sion. For example, when converting a temperature from degrees centigrade to
degrees Fahrenheit, an approximation is to multiply the temperature in degrees
centigrade by 5/9 and then add 32. Assuming that the variable c contains a temperature in degrees centigrade, you could write the following in Python:

```
>>> c = 20
>>> f = (c * 9/5) + 32
>>> f
68.0
```

The parentheses are not strictly necessary here because MicroPython will automatically perform multiplication and division before it does addition. But including them does make the code clearer.

Strings

Computers are really good at numbers. After all, this is what they were originally
created for. However, in addition to doing math, computers often need to be able
to use text. Most often this is to be able to display messages that we can read. In
computer-speak, bits of text are called *strings*. You can give a variable a string
value, just like numbers. Type the following into the REPL.

```
>>> s = "Hello"
>>> s
'Hello'
```

So first we assign a value of Hello to the variable s. The quotation marks around
"Hello" tell Python that this is a string and not, say, the name of some variable.
You can use either single or double quotation marks, but they must match. We
then check that s does contain "Hello".

Rather like adding numbers, you can also join strings together (called *concatenation*). Try the following in the REPL:

```
>>> s1 = "Hello"
>>> s2 = "World"
>>> s = s1 + s2
>>> s
'HelloWorld'
```

This might not be quite what we were expecting. Our strings have been joined together, but it would be better with a space between the words. Let's fix this:

```
>>> s = s1 + " " + s2
>>> s
'Hello World'
```

Converting Numbers to Strings

Try the following example, which tries to combine a string with a number:

```
>>> x = 2
>>> s = "answer is:"
>>> s + x
Traceback (most recent call last):
  File "<stdin>", line 1, in <module>
TypeError: must be str, not int
```

You assigned an int value of 2 to x and a string "answer is:" to a variable s. However, when you try to add the number onto the end of the string, you get an error message. The key part of this error message is the last line `TypeError: must be str, not int`. This is saying that there is an error with the type of the variable and that you are trying to add an int to a str (*string*).

To get around this problem, you need to convert the integer to a string before trying to add it, like this:

```
>>> s + str(x)
'answer is:2'
```

The command str is what's called a *function*, and you will find out a lot more about functions in Chapter 4. For now, though, all you need to know is that if you place a number or a variable containing a number inside the (and) after str, the resulting string version of the number can be added to another string.

Programs

The commands that we have been typing into the REPL are single-line commands that just do one thing. You can see how typing them one after the other leads to their being run (or *executed*) one after the other. For example, revisiting this example:

```
>>> s1 = "Hello"
>>> s2 = "World"
>>> s = s1 + s2
>>> s
```

The four command lines are executed one after the other as fast as you can type them. If you were to put all four lines into a file and then tell the micro:bit to execute the whole file, you would have written a *program*.

Let's try this now, but first we need to make one slight change. When you use the REPL command line and simply type the name of a variable, the REPL will display the value in that variable. However, when you run the same commands as a program, you must use the `print` function for any value that you want to display. So click on New in Mu to create a new program, and then type the following lines into the window (Figure 3-2):

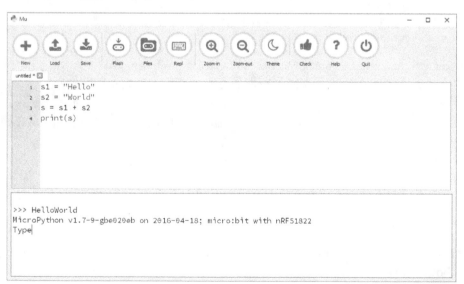

Figure 3-2   *A very small program.*

```
s1 = "Hello"
s2 = "World"
s = s1 + s2
print(s)
```

Click on the REPL button so that we can see the output from the micro:bit, and then click on Flash to upload the program. When the program has uploaded, you will see the text "HelloWorld" in the output of the REPL. Your program has worked; it joined two strings together and printed out the result. You can now easily change the program so that instead of using print to display the string, it scrolls the message across the micro:bit's display.

You can either change the code in the Mu editor so that it looks like this:

```
from microbit import *
s1 = "Hello"
s2 = "World"
s = s1 + s2
display.scroll(s)
```

Or you can go and find the program files that you downloaded for this book back in Chapter 2 and open the program ch03_strings.py. Now when you run the program, the message "HelloWorld" will scroll across the screen just once.

The micro:bit has run each of these command lines in turn and then finished by showing s on the display. Having run all its commands, the micro:bit has nothing further to do. The program has ended.

Looping Forever

Generally speaking, you don't want the program running on a micro:bit to end. It's not like running an app on your computer that you can just quit when you have finished using it. A program on a micro:bit will basically run until you unplug it. It might be waiting for button presses, reading a sensor, or displaying information, but it will be running. This is why you find something called a while loop at the end of most micro:bit programs.

Either load program ch03_loops.py or type the following code into Mu:

```
from microbit import *
while True:
    display.scroll("Hello")
    sleep(5000)
    display.scroll("Bye")
    sleep(5000)
```

Before we look at what this code does, an important point about indentation needs to be made. You will notice that after the while True: line, the remaining lines are all indented by four spaces. This indicates that those lines belong inside the while command. Python insists that they line up exactly (or you will get an error message), and the convention is to use four spaces of indentation. Helpfully, in Mu, when you press the TAB key, you get four spaces inserted for you.

What happens when you upload (*flash*) this program onto the micro:bit is that the messages "Hello" and "Bye" appear in turn on the display and repeat until you unplug the micro:bit or upload a different program.

Here's how the code works. The command while True: tells the micro:bit to repeat all the indented lines (after the colon on the end of the line) forever. The "Hello" message is then displayed. The line sleep(5000) tells the micro:bit to do nothing (*sleep*) for 5,000 milliseconds. A milliscond is 1/1,000 of a second, so 5,000 of them is 5 seconds. The "Bye" message is displayed, followed by another 5-second sleep, and then the cycle begins again.

The line:

```
display.scroll("Hello")
```

requires some special attention. The variable display is called an *object*, and it mirrors what is going on with the physical hardware of the micro:bit—in this case the LED display. The function scroll() is a special kind of function called a *method* because it belongs to the display object. The dot between the object and its method is how you tell Python to use the method.

If this all seems a bit mysterious, don't worry; Chapter 7 is all about objects and their relatives: classes and methods.

for Loops

In this section, you will learn about for loops. This means telling Python to do things a number of times rather than just once. In the following example, you will need to enter more than one line of Python using the REPL.

```
for x in range(1, 10):
```

When you hit RETURN and go to the second line, you will notice that Python is waiting. It has not immediately run what you have typed because it knows that you have not finished yet. The colon (:) at the end of the line means that there is more to do.

For the second line, type:

```
print(x)
```

The REPL will automatically indent the second line for you.

To get the two-line program to actually run, press BACKSPACE and then ENTER after the second line is entered. Pressing BACKSPACE deletes the indentation so that the REPL knows that you have finished. Thus:

```
>>> for x in range(1, 10):
...        print(x)
1
2
3
4
5
6
7
8
9
>>>
```

This program has printed out the numbers between 1 and 9 rather than 1 and 10. The range command has an exclusive end point—that is, it doesn't include the last number in the range, but it does include the first.

There is some punctuation here that needs a little explaining. The parentheses are used to contain what are called *parameters*. In this case, range has two parameters from (1) and to (10) separated by a comma.

The `for in` command has two parts. After the word `for`, there must be a variable name. This variable will be assigned a new value each time around the loop. So the first time it will be 1, the next time 2, and so on. After the word `in`, Python expects to see something that works out to be a list of things. In this case, that list of things is a sequence of the numbers between 1 and 9.

The `print` command also takes an argument that displays it in the REPL area of Mu. Each time around the loop, the next value of x will be printed out.

---

ifs and elses

Most programs are more than just a simple list of commands to be run over and over again. They can make decisions and do some commands only if certain circumstances apply. They do this using a special Python word called `if` and its optional counterpart, `else`.

To illustrate this, we are going to make your micro:bit respond to presses of button A by displaying a message. Either load program `ch03_button.py` or type in the following code:

```
from microbit import *

while True:
    if button_a.was_pressed():
        display.scroll("Please don't press this button again")
```

Inside the `while` loop we have an `if`. Immediately following the `if` is a *condition* that must be either `True` or `False` and then a colon. In this case, the condition is that `button_a.was_pressed()`. `button_a` is another object like `display` and is an object that mirrors what is going on with the physical button A. There is a counterpart for button B called (you guessed it) `button_b`. Whereas the *scroll* method of `display` instructs the micro:bit's processor to scroll a message across the screen (in other words, to do something), the `was_pressed` method asks a question of the hardware ("Were you pressed?"). This answer can only be yes or no, which in Python is represented as `True` or `False`.

Notice how the last line is double indented, one indent to put the code inside the `if` and another because the `if` is inside the `while`.

It is not uncommon to want your code to do one thing if the condition is `True` and another if it is `False`. Let's modify the preceding example. You can find the modified code in `ch03_button_else.py`.

```
from microbit import *

while True:
    if button_a.was_pressed():
        display.scroll("Please don't press this button again"")
    else:
        display.scroll("Press button A")
```

The lines after `else` will only be run if the condition of the `if` was `False`. So now the micro:bit will repeatedly display the message `Press button A` until such time as you do, and then it will tell you not to do it again.

Sometimes you need more than just `if` and `else`. You actually need some conditions in between. In that case, you can use `if` followed by the word `elif` as many times as you need, optionally followed by an `else`. The following example illustrates this:

```
if x < 5:
    print("x is small")
elif x < 10:
    print("x is medium")
elif x < 100:
    print("x is large")
else:
    print("x must be enormous")
```

More on `while`

Earlier we saw how `while True:` is used to keep the micro:bit running indefinitely. The `while` command is actually more flexible than this. Whereas `if` will do something once if its condition is `True`, `while` will keep doing the things below it repeatedly while its condition is `True`.

By using `while True:`, the condition of the `while` will always be `True`, so the `while` loop will never finish. You can also put a condition after the `while` that could be `True` or `False`, as in the example of ch03_while.py:

```
from microbit import *

while not button_a.was_pressed():
    display.scroll("Press A")
display.scroll("Terminating!")
```

Flash the program and try it out. The while loop ensures that as long as button A is not pressed, the message "Press A" will be displayed. When the button is pressed, the condition for staying inside the loop is no longer True (note use of not), so the last line of the program that is outside of the loop will run, and then the program will quit. Your micro:bit will look like it has died. Don't worry, it will revive when you upload another program.

Timer Example

In the next few chapters, you will gradually build a kitchen timer project using the micro:bit. Using what you have learned so far, you can make a start on the part of the timer that will let you set the number of minutes to time for by pressing button A. When the number of minutes exceeds 10, you will set it back to 1. You can find this code in ch03_timer_01.py.

```
from microbit import *

mins = 1

while True:
    if button_a.was_pressed():
        mins += 1
        if mins > 10:
            mins = 1
    display.scroll(str(mins))
```

The variable mins is used to remember the current number of minutes. This is initially set to 1. This has to happen outside the while loop; otherwise, mins would just keep getting reset back to 1 each time around the loop.

Inside the loop, the if checks for a button press, and if there is one, it adds 1 to mins. This might make mins greater than 10, so we have a second if inside the first that checks for this and, if it's the case, sets mins back to 1.

In this case, the condition for the if is that mins > 10. This means "if the value of the variable mins is greater than 10." You can do all sorts of comparisons here: you can compare whether things are equal in value (==), not equal (!=), less than (<), greater than (>), less than or equal to (<=), or greater than or equal to (>=). These comparisons are like questions and always give a result of either True or False (mins is greater than 10 or it isn't—there is no middle ground).

Note that the == (double equals) comparison is used a lot, and a common mistake is to accidentally use a single equals sign in comparisons. The single equals sign is only ever used to set the value of a variable.

Conditions can be combined into more complicated forms. For example, if for some bizarre reason you only wanted the number of minutes to be increased if both buttons were pressed, you could modify the program to look like this:

```
from microbit import *

mins = 1

while True:
    if button_a.was_pressed() and button_b.was_pressed():
        mins += 1
        if mins > 10:
            mins = 1
    display.scroll(str(mins))
```

Try making the change, and then upload the code. You may have noticed that buttons A and B do not have to be actually down at the same time; you can also press one after the other quickly. This is so because the button object remembers when it has been pressed for the next time that the question was_pressed() is asked.

Summary

This chapter has covered quite a lot of ground in terms of basic programming concepts. If you are new to programming, it can take time to make sense of these ideas, so you may want to play with the code examples so far. Alter them, upload them, and see what happens. Experimenting is a great way to learn. If you mess up the programs, it doesn't matter; you can just download them again. In Chapter 4, you will learn all about one of the most important features of the Python language—functions.

4

Functions

You have already used built-in *functions* when you have done such things as str(x), which converts a number to a string. You have also used display.scroll(). scroll is a special kind of function that belongs to the display object. Such functions are called *methods*. The scroll method will "call" other methods and functions to turn LEDs on and off and make the display do what it does. In addition to all the built-in functions and methods in Python, you can write your own functions. This helps to keep your code neat and easy to follow.

What Are Functions?

You can think of functions as commands that you can use in your programs, such as str (convert something to a string) or sleep (wait a while). These might be built-in functions, or they might be functions that you have written yourself. Writing your own functions allows you to have your program in small chunks of code so that it's easier to understand how the program works and to fix things when they don't work. It also means that you can use the function over and over again without having to repeat chunks of code.

These chunks of code (functions) are given a name when you make them, and the name has the same rules as when you create a variable. That is, the name must start with a letter and not contain any spaces. Words are connected by an underscore, so typical function names might be sleep, print, or, for a home-made function that you will see in a moment, display_mins. Whereas the names of variables are normally nouns (e.g., message and mins), the names of functions are often verbs (e.g., sleep, print, and display_mins).

Parameters

Functions can be built to have *parameters*. For example, the built-in function `sleep` has a parameter that is the number of milliseconds for the program to sleep. The function `sleep` expects to receive this parameter whenever `sleep` is called. In the case of `sleep`, it expects this parameter to be a number. For example:

```
sleep(5000)
```

If you were to `pass` a string to `sleep` as its parameter, you would get an error. When calling a function, the parameter always comes after the method name and is enclosed in parentheses. If a function has more than one parameter, the parameters are separated by commas.

The time has come for you to write your own function in our timer example. So load the example `ch04_timer_02.py`. The revised program is shown in Figure 4-1.

```
 1  from microbit import *
 2
 3  mins = 1
 4
 5  def display_mins(m):          Parameter
 6      message = str(m)
 7      if m == 1:
 8          message += " min"      ──── Function definition
 9      else:
10          message += " mins"
11      display.scroll(message)
12
13  while True:
14      if button_a.was_pressed():
15          mins += 1
16          if mins > 10:
17              mins = 1
18      display_mins(mins)         ──── Function call
19
```

Figure 4-1 *Function definition and calling.*

Flash the program onto your micro:bit, and notice that now it displays either `min` or `mins` after the number of minutes to which the timer is set. The next thing to notice about the code is that now there is a block of code that starts with def after the mins variable. The word def marks the start of a method definition. This is where our custom function called display_mins is defined.

And here is a really important point: although the function is defined here, it will not actually do anything until it is *called*. This calling of the function only happens right at the end of the while loop in the last line of the program.

In the same way as if, else, and while, all the lines that belong to the function definition must be indented.

After the word def is the name of the function and then its parameters between parentheses. In this case, there is just the one parameter called m. m is short for mins, and we could have called m by the name mins like the variable. But I have used a different name here to make it clearer how functions work and the relationship between m inside the function and mins outside it.

In addition to being a parameter, m is also a special type of variable called a *local variable*. It is local in the sense that it can only be used within the function of which it is a parameter (in this case, display_mins). The value of m is set when the function is called, so jumping for a moment to the end of the program, we have:

```
display_mins(mins)
```

This line of code calls (runs) the function display_mins, specifying that it will supply mins as the parameter to the function. Python will take whatever value is in mins and assign it to m within display_mins. The lines of code inside display_mins will then be run in turn until the end of the function is reached, at which point Python will continue with any lines that come after the call to the function. In this case, there is no line of code after the call. It's the last line, but because it is inside the loop, the first line inside the loop will run again.

Returning for a moment to the lines of code in the function (called the function's *body*), you can see that the variable message is defined inside the function and is given a value of m converted into a string. This variable m, although not a parameter like m, is still a local variable. That is, the variable message will only exist inside the function while the function is actually running. So a new message is created each time that the display_mins function is called and is then automatically destroyed after the function finishes running. This may seem inefficient, but actually microprocessors are designed to be able to do this kind of thing really quickly.

The if on line 7 checks to see if the number of minutes m is 1, and if it is, it adds the string "min" to message. Note the use of +=, which adds the string "min" onto the end of message. If m isn't 1, then "mins" (plural) is added to the message. Finally, display.scroll is called to display the message.

Global Variables

Earlier I described m and message as local variables. The variable mins sits at the top of the program and is accessible from anywhere within the code. This type of variable is called a *global variable*. The discussion about just how bad it is to use global variables takes on almost religious overtones. There is no doubt that over-use of global variables does lead to unmanageable code, but if you are writing a few dozen lines of Python for a micro:bit, it's really not the same problem as the thousands or millions of lines of code in a large program.

If you wanted to, you could just use the global variable mins throughout the program. You might like to load up ch04_timer_02_globals.py and verify that it does work the same as the previous version.

```
from microbit import *

mins = 1

def display_mins():
    message = str(mins)
    if mins == 1:
        message += " min"
    else:
        message += " mins"
    display.scroll(message)

while True:
    if button_a.was_pressed():
        mins += 1
        if mins > 10:
            mins = 1
    display_mins()
```

The differences between this and the previous version are highlighted in bold. First, the parameter m has been removed, so display_mins no longer has any parameters. Now `display_mins` uses the global variable `mins` directly rather than having `mins` passed to it.

This version works just fine in this simple example, but it does mean that you can now only use `display_mins` to display the value in the global variable `mins`. So if we had another global variable called something else, we couldn't use `display_mins` to display it. By giving it a parameter, we make it more flexible.

Return Values

The `display_mins` function does a task rather than calculate a value. It's not like the built-in function `str`, which takes a number as a parameter and converts it into a string. Functions like `str` are said to *return a value*. This allows you to write things like this:

```
message = str(123)
```

which gives the variable `message` the value returned by `str` (in this case, the string `"123"`).

You can write your own functions that return a value using the word `return`. For example, we can modify our timer example so that the code to add `min` or `mins` to the end of the number of minutes is contained in its own function. You can find this modified version in `ch04_timer_03.py`.

Most of the code is the same, but `display_mins` has been changed, and a new function, `pluralize`, has been added. Let's look at `pluralize` first.

```
def pluralize(word, x):
    if x == 1:
        return word
    else:
        return word + "s"
```

This function has two parameters: `word`, which is a text string, and `x`, which is an integer. If `x` is 1, then the function *returns* the parameter `word` unchanged. However, if `x` is not 1, then the value of `word` plus an `"s"` is returned. Once the word `return` is encountered, nothing further happens in the function. So in addition to returning a value, `return` also ends the function, continuing with the main program.

We can now use the `pluralize` function in the `display_mins` function by changing `display_mins` to:

```
def display_mins(m):
    display.scroll(str(m) + pluralize("min", m))
```

The body of display_mins is now just a single line that contains two calls to other functions. First, it calls the built-in function str to convert the number into a string, and then it calls pluralize to add an s if necessary. The results of these two function calls are combined into a single string using a plus sign, and the result is scrolled across the display.

Advanced Parameters

One neat feature of Python is the ability to use optional parameters with default values. This allows you to hide away rarely used parameters while retaining the flexibility of the function. For example, the following function *increases* the number supplied as an argument. By default, it will increase it by just one.

```
def increase(x, amount=1):
    return x + amount

print(increase(5)) # prints 6
print(increase(5, 10)) # prints 15
```

Summary

Functions are a great way to break down your programs into more manageable chunks, and functions can also often be reused, so you can take a function that you wrote for one project and use it in another. You will meet *methods*, which are a special type of function, in Chapter 7. We are now going to park the timer example for a while and return to it in Chapter 6, where we will make it a whole lot more complex and finish the project.

5

Lists and Dictionaries

There are two ways of structuring data that are extremely useful and are found in most programming languages. One is a way of representing lists of things (called *Lists* in Python), and the other links data with a key (called *Dictionaries* in Python). In this chapter, you will learn all about Lists and Dictionaries and also build on what we have learned so far to write some more programs. To distinguish Python Lists and Dictionaries from the everyday use of those words, I will capitalize their first letter.

Lists

To get to know Python Lists, we will use the example program of a message board that allows us to select one message to be displayed from a number of predefined messages. You could, for example, make this into a physical project that you could leave on your door to express messages to other people in your house without having to resort to anything so crude as pen and paper.

Message Board Example

Rather than dealing with Lists in an abstract way, let's jump right into an example, and then we can deal with the details of how you can use Lists in more detail later. This is a really simple project to make. You can use a battery power source for your micro:bit, or if you are going to place it close to an AC outlet, you can use a USB power lead. You will also need some way of fixing it to your door. Figure 5-1 shows my approach. The battery is a USB rechargeable battery pack that costs less

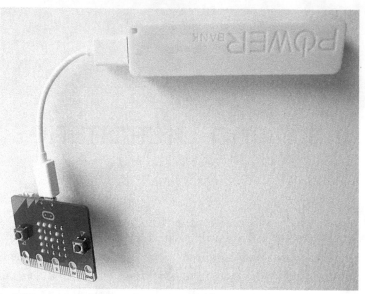

Figure 5-1 *Message board example.*

than $5 and will power a micro:bit for around 24 hours before needing to be recharged. It's stuck to the door with adhesive pads, and the micro:bit is attached by a short USB power lead.

The full code for this project can be found in ch05_message_board.py. The Listing for this program is broken into chunks. The first thing to note is the messages List.

```python
from microbit import *

messages = [
  "-",
  "I've gone out",
  "Do not disturb",
  "I'm at the shops, message me if you can think of anything we need",
  "I'm in the garden",
  "Give me a call",
  "I'm in the bath, a cup of tea would be nice"
]

message_index = 0
```

The variable messages is similar to other variables that you have used, but instead of having a value that is a single number or string, messages is given a value that is a list of strings. Python knows that it's a List because the values for the List are enclosed in square brackets. Each element of the List is separated by a comma, and to make the messages easier to read, each is on its own line indented by four spaces. The first message is a simple "-" to indicate no message. You can replace my messages with messages of your own, and you do not have to have the same number of messages.

The global variable message_index indicates the index position in the List of the message to be displayed. This is the number of the string in the List to be displayed. An important point is that this is initially set to 0 rather than 1. This is so because index positions in Python start at 0. So the first element in the List is 0, the second 1, the third 2, and so on.

```
def show_current_message():
    display.scroll(messages[message_index], wait=False, loop=True)
```

The function show_current_message calls display.scroll. The first parameter in the message to be displayed is messages[message_index]. The square bracket after the List name containing an index number allows you to fetch one particular element from the List. The two optional parameters to display.scroll tell the micro:bit not to wait until the message display has finished before carrying on with the rest of the program, and the loop parameter tells it to keep displaying the message repeatedly. Immediately after show_current_message is defined, it is also run, so the first message of "-" is displayed.

Next, we come to the main while loop. This checks for a button press on button A, and if it finds one, it adds one to message_index. To make sure that we don't run off the end of the List and cause a *range error*, the built-in function len is used to find the length of the List and compare it with message_index. If message_index is greater to or equal to the length of the messages List, then message_index is set back to 0.

Finally, and still only if button A is pressed, the message to be displayed is changed to the message from the List with the index position message_index.

```
show_current_message()

while True:
    if button_a.was_pressed():
        message_index += 1
```

```
if message_index >= len(messages):
    message_index = 0
show_current_message()
```

The Lowdown on Lists

This is quite a high-speed tour of the program, so now let's go into more detail about how Lists work and what you can do with them. There is also a section in Appendix A that provides a reference for using Lists.

Accessing Elements of a List

As you saw earlier, to fetch a particular element of a List, you can put the index number of the element that you want after it inside square brackets. Start the REPL from Mu, and then try out the following commands from your micro:bit's command line:

```
>>> x = [1, "aaa", 23, 34, 77]
>>> x[0]
1
>>> x[1]
'aaa'
>>> x[5]
Traceback (most recent call last):
  File "<stdin>", line 1, in <module>
IndexError: list index out of range
>>>
```

Here you have created a List in a variable called x that has five elements, which are a mixture of four integers and one string. When you come to try and access the element with index 5, you get an error message telling you that the index is out of range.

In addition to being able to get a particular element, you can also set it. Now try entering the following lines:

```
>>> x
[1, 'aaa', 23, 34, 77]
>>> x[2] = "bbb"
>>> x
[1, 'aaa', 'bbb', 34, 77]
>>>
```

Entering just x on its own allows you to see the contents of the List. You then set list element 2 (the third element) to "bbb". Displaying the contents of the list again confirms the change.

In addition to being able to get individual elements of a List, you can also get a section of a List. For example, if you wanted the second and third elements of the List, you would type:

```
>>> x[1:3]
['aaa', 'bbb']
```

This is a little confusing because you might expect the number after the colon to be 2, but actually the end point of the range is exclusive. That is, it is one more than the last index.

Adding to a List

You can add elements to the end of a List using append, as illustrated in the next piece of code. Note that we are starting with a fresh List.

```
>>> x = [1, "aaa", 23, 34, 77]
>>> x.append(99)
>>> x
[1, 'aaa', 23, 34, 77, 99]
```

If you want to add a new element at a different position in the List, you can use insert.

```
>>> x.insert(1, 11)
>>> x
[1, 11, 'aaa', 23, 34, 77, 99]
```

The first parameter of insert is the position in the array where the new element should go, and the second parameter is the element to insert (in this case, the integer 11).

Deleting from a List

The pop command allows you to remove an element from the List. By default, pop removes the last element from the List and returns it.

```
>>> x
[1, 'aaa', 23, 34, 77, 99]
```

```
>>> x.pop()
99
>>> x
[1, 'aaa', 23, 34, 77]
>>>
```

You can also "pop" a specific element from the List by using pop with a parameter
of the position of the element to be popped.

```
>>> x
[1, 'aaa', 23, 34, 77]
>>> x.pop(2)
23
```

Joining Lists Together

In the same way that you can join together strings using the plus sign (+), you can
also use plus to join to arrays like this:

```
>>> x1 = [1, 2, 3]
>>> x2 = [4, 5, 6]
>>> x1 + x2
[1, 2, 3, 4, 5, 6]
```

Note that this does not alter either 11 or 12; it returns a new List that is a combi-
nation of the two.

Strings as a List of Characters

Python lets you do many of the things that you can do to a List to strings. After all,
a string is actually a list of characters. So you can do things like this to access a
particular character of a string:

```
>>> s = "Hello World"
>>> s[0]
'H'
```

You cannot, however, modify a particular character in a string. If you try, you will
get an error message like this:

```
>>> s[0] = "h"
Traceback (most recent call last):
  File "<stdin>", line 1, in <module>
TypeError: 'str' object does not support item assignment
```

You can also splice off sections of a string (called *substrings*) like this:

```
>>> s = "Hello World"
>>> s[0:5]
'Hello'
```

The function `len` works on both Lists and strings:

```
>>> s = "Hello World"
>>> len(s)
11
```

You will find a handy reference for using strings in Appendix A.

Dictionaries

Whereas you access elements of a List by their position, elements of a Dictionary are accessed by a *key* that can be a string or a number or pretty much anything. A value is associated with each key, so to find an element in a dictionary, you supply a key, and the Dictionary responds with the appropriate value.

An Example

The following mini-project (Figure 5-2) displays a different symbol on the display to represent the last movement gesture detected by the micro:bit.

You will learn more about gestures in Chapter 9, but the basic idea is that the micro:bit can use its accelerometer to detect the orientation of the board (face up, face down, tilted left, tilted right, tilted forward, or tilted backward). The program displays a different symbol for each type of gesture.

One way to write this program would be to have a whole load of `if`s. Actually, there is nothing wrong with that, but the solution here has the advantage that it is easier to change the symbols to display because they are all at the top of the program in a Dictionary rather than being spread about the program. You can find the program in `ch05_gesture_detector.py`.

```
from microbit import *

gesture_chars = {
    "up" : "^",
    "down" : "v",
```

```
        "left" : "<",
        "right" : ">",
        "face up" : "+",
        "face down" : "O",
}

while True:
        gesture = accelerometer.current_gesture()
        if len(gesture) > 0:
            display.show(gesture_chars[gesture])
```

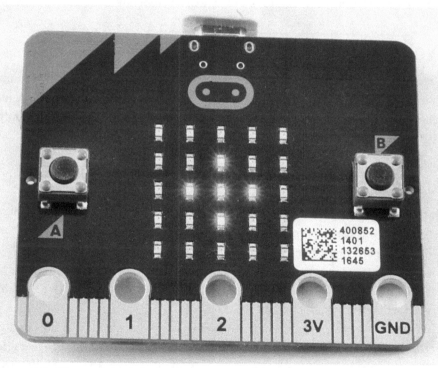

Figure 5-2 *The gesture-detector project.*

Whereas a List uses square brackets around its contents, a Dictionary uses curly braces ({ and }). Within the curly braces, each element is in two parts: the key before the colon and the value after it. You do not have to put each key/value pair on a new line, but I have done so here to make it easier for you to see what's going on.

Looking at the Dictionary `gesture_chars`, you can see that the key of "up" has a value of "^" and that of "down" has a value of "v", and so on. The while loop first gets the current gesture, which will be a text string, and then, as long as the string is not empty (length is greater than 0), it looks up the character to "show" using the same square-bracket syntax that you use with a List, except that in this case you have a key (in this case, a string) inside the square brackets rather than a position index.

Upload the program, and try tilting and shaking the micro:bit to see the display change.

Dictionaries in Detail

Let's look in more detail at how to use Dictionaries. To get the hang of Dictionaries, open the REPL, and be prepared to try some experiments. You will also find a section in Appendix A that gives more information on using Dictionaries.

Accessing Elements of a Dictionary

Try this example in the REPL:

```
>>> d = {1:"one", 5:"five", 7:"seven"}
```

This line creates a new Dictionary with three elements. Each element has an integer as its key and a string as its value.

To retrieve one of the values, you must know its key. The following line will return the element with a key of 5:

```
>>> d[5]
'five'
```

If you attempt to access an element of the Dictionary using a key that does not exist, you will get an error message:

```
>>> d[0]
Traceback (most recent call last):
  File "<stdin>", line 1, in <module>
KeyError: 0
```

The word `KeyError` tells you that the key has not been found. This can be inconvenient, especially if you just want to see if the key exists in a Dictionary. To avoid such errors, use `get` rather than the square-bracket notation.

```
>>> d = {1:"one", 5:"five", 7:"seven"}
>>> d.get(0)
>>> d.get(0) == None
True
>>> d.get(1)
'one'
```

Now, when you try to get the value of a key that doesn't exist, you get the special value of None. None is used in Python to signify the absence of something.

Adding Items to a Dictionary

You can add items to a Dictionary like this:

```
>>> d = {1:"one", 5:"five", 7:"seven"}
>>> d[2] = "two"
>>> d
{1: 'one', 5: 'five', 7: 'seven', 2: 'two'}
```

To remove an element from the Dictionary, you can use pop. The parameter to pop is the key of the element to remove, and pop modifies the Dictionary and then returns the value for that key.

```
>>> d
{1: 'one', 5: 'five', 7: 'seven', 2: 'two'}
>>> d.pop(2)
'two'
>>> d
{1: 'one', 5: 'five', 7: 'seven'}
```

Summary

In this chapter, you have learned more about some of the fundamentals of Python. In Chapter 6, you will return to the timer project and also look at some more advanced programming techniques, as well as how to go about dealing with complex designs.

6

Writing Your
Own Programs

Writing Software

Writing programs for a device such as the micro:bit is actually a little different from the job of writing software for a "real" computer. When you are writing an app for a computer or phone, you will be writing a user interface that uses a large-resolution display (compared with the micro:bit) and have the benefit of a keyboard and mouse or touch screen to use as an input device. Also, an app for a computer or phone will be running on a processor hundreds of times faster with thousands of times more memory.

One of the joys of writing for a micro:bit is that by necessity the programs need to be short and sweet and work efficiently. There is no room for "bloat-ware." Software written for small devices such as the micro:bit is called *embedded software* and (influenced by electronics engineers) comes with its own culture that is a little different from large-scale software development, although there are many good tricks to be taken from "big" software.

Keep It Simple

In particular, when you first start programming, it is better to start small with simple examples and gradually build on them rather than to type in hundreds of lines of code and then be faced with a mountain of errors and a confusing program that will then consume days of your time to fix. It takes time to track down and fix problems. Although it might seem like you waste a lot of time waiting while your latest version of the code is flashed onto the micro:bit, believe me, it's much quicker to make small changes between each flashing than it is to track down tricky bugs.

Spikes and the REPL

When learning something new—perhaps exploring how the "sound" library works—try things out in the REPL. Reading the documentation is one thing, but actually trying things makes everything a lot more concrete. If you end up stringing a few lines together in the REPL, then you can always copy and paste them into your program later.

Similarly, professional programmers will often write *spike solutions*, small programs to test out a tiny part of a much bigger system. Once the *spike* has done its job of proving that something works, it is generally thrown away, although bits of it might be copied and pasted into the real program.

Versioning

In professional software development, every file that a programmer works on will be kept in a version management system, and whenever a change is made to a file of code that completes a small task of work, a new version of the file will be created. The version management system allows the programmer to, at any time, revert to how the code looked in an earlier version.

Version management is essential in professional programming but is not really necessary for the small programs usually developed by a single person for the micro:bit. However, it is a really good idea just to make a copy of your micro:bit program file before you make any major changes. You can do this in your computer's file system, and a common convention is to make a copy of the file with _org (for "original") after the program's name but before the .py. In this way, if you mess up your program, you can just delete the broken program and rename the last working version, removing the _org to reinstate it.

Another option, if you do mess up your program, is Mu's undo feature. This can be really useful. Mu remembers the recent changes you have made. On Windows and Linux computers, to undo your latest change, press CTRL-Z, and on a Mac, use CMD-Z.

Comments

Any line in Python that has a # indicates that the rest of the line after the # is a *comment*. Comments are ignored by MicroPython. They are just messages to anyone reading the program (including you). Good reasons for adding comments to your code include

- To explain what the program does and maybe who wrote it
- To explain how to tweak parts of the program that someone might need to change
- To insert before a function, where the purpose of the function requires more explanation than just the function name itself
- To explain anything that is tricky to understand and you can't make obvious by refactoring (see later section)
- As an aid to beginners who are learning to program

Aside from the "aid to beginners" reason, there is no point in commenting on the glaringly obvious. For example, writing the following is not only pointless, but it can also actually be misleading:

```
x += 1   # add 1 to x
```

The code is already *self-documenting*, so it doesn't need an extra comment. The reason that such comments are bad is that you now have two things to maintain and keep in step. So, if you later find that the code is supposed to subtract 1 rather than add 1, you have to change both comment and code. Often this double change doesn't happen, leaving inconsistency. People who program for a living will generally mistrust comments and "let the code speak."

Usually, when programming, comments have no effect on the size of the final running program. However, MicroPython is a little different in that the whole text of the program (comments and all) is stored in the micro:bit's flash memory. So, if when you try to flash your program onto a micro:bit you get the error message, "Unable to flash. Your script is too long," try removing the comments.

Refactoring

Good programmers are not judged by how many lines of code they write a day but by how few they need to write to get the job done. Good programmers have a

need to keep things neat and tidy, and when they find code that is overcompli-
cated or repeats itself, they will *refactor* it, simplifying it and thereby reducing its
size and making it easier to work on.

The acronym *DRY* (Don't repeat yourself) means that where you find yourself
repeating the same lines of code more than once in the same program, you probably need to refactor those lines into their own function. For example, the following version of `ch5_message_board.py` will work just fine but could be improved:

```python
from microbit import *

messages = [
    "-",
    "I've gone out",
    "Do not disturb",
    "I'm at the shops, message me if you can think of anything we need",
    "I'm in the garden",
    "Give me a call",
    "I'm in the bath, a cup of tea would be nice"
]

message_index = 0

# Display the current message
display.scroll(messages[message_index], wait=False, loop=True)

while True:
    if button_a.was_pressed():
        message_index += 1
        if message_index >= len(messages):
            message_index = 0
        # Display the current message
        display.scroll(messages[message_index], wait=False, loop=True)
```

The two fairly long highlighted lines are identical. They both display the current
message, but this needs to be done in these two places in the program. The fact
that both lines are preceded by a comment saying what the next line does is a dead
giveaway that this line should be made into its own function so that the comments
become unnecessary because the meaning is there in the function name.

Here is the DRY version of the relevant part of the program:

```
def show_current_message():
    display.scroll(messages[message_index], wait=False, loop=True)

show_current_message()

while True:
    if button_a.was_pressed():
        message_index += 1
        if message_index >= len(messages):
            message_index = 0
        show_current_message()
```

To illustrate why DRY code is important, imagine that you want to make the message scroll more quickly by using the delay optional parameter to scroll. Since refactoring the code, all you need to do is alter the one line inside show_ current_message() to be

```
display.scroll(messages[message_index], wait=False, loop=True,
        delay=100)
```

State Machine Design

The timer example that we started back in Chapter 3 actually gets surprisingly complicated when you start to think about how it should behave. What happens when you press the buttons is different depending on whether you are setting the time on the timer, or the timer is counting down, or the timer is buzzing to indicate that time is up. Programs for devices such as the micro:bit often need to make use of the idea of the program being in different modes or *states*, and a useful way to plan out exactly how the program should work and move between these modes is called a *state machine*. Actually, this concept, which has come from mathematics, is more properly called *finite state machines* (FSMs). But don't worry, despite the very formal name, it's actually pretty easy to draw a state machine diagram as an aid to writing your code.

The first step is to identify the states (think *modes*) that the timer can be in. It can be in one of the following three states:

- SET—setting the number of minutes from which the timer will start

- RUN—counting down

- ALARM—sounding the alarm because the countdown has finished

By convention, constants such as SET, RUN, and ALARM are written in uppercase letters.

You now need to plan out just what will cause your program to change from one state to another and anything that may happen as that transition occurs. The best way to do this is as a diagram like the one in Figure 6-1.

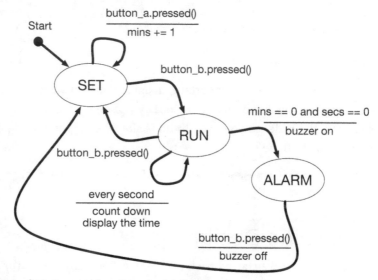

Figure 6-1 *A state machine diagram for the timer example.*

The first arrow from Start indicates that when the program runs; SET should be the first state that it enters. Once in the SET state, pressing button A will increase the number of minutes by one. This is indicated by a piece of text above the line (button_a.pressed())—this is the condition. Below the line is the "action" to carry out (mins += 1), and the arrow bends all the way back to SET, indicating that after button A has been pressed, we still stay in the SET state. Here mins += 1 is actually shorthand for the whole adding of 1 to the minutes and wrapping around back to 1 when you get to 10 that we explored in Chapter 3.

If we are in the SET state and button B is pressed, then we change state to RUN, and we don't need to do anything else when we change state, so there is no need for a line or an action. From the RUN state, pressing button B again will take you straight back to SET. Also, while in the RUN state, the time will count down every second and be displayed. When the numbers of minutes and seconds both reach zero, the buzzer will be turned on, and we will enter the ALARM state, from which we will return to the SET state, turning the buzzer off, when button B is pressed.

Revisiting the Timer Example

It's time to revisit the code for the timer example and use the state machine diagram of Figure 6-1 to rewrite the code. So disregard what we have done so far in the timer example; we are going to start again.

Rather than trying to type in all the following code, I suggest loading the program ch06_timer_03 to program ch06_timer_06 as they are referenced in the following discussion.

A State Machine Framework

Let's start with the basic template for representing the SET, RUN, and ALARM states of our program. The following code isn't a complete program, so don't try to run it.

```
SET, RUN, ALARM = range(3)
state = SET
```

The first line above assigns the values 0, 1, and 2 to SET, RUN, and ALARM. Each state has a unique number, but we don't need to use the number for anything other than telling the states apart because we will always refer to the state by its variable name to make the code easier to read.

The variable state is a global variable used by the program to keep track of the current state. To help keep the state in manageable units, three functions will be used to "handle" what goes on in each state. These are empty at the moment, but we can start to fill them out.

```
def handle_set_state():

def handle_run_state():

def handle_alarm_state():
```

The main while loop is a sequence of if/elif commands that call the appropriate handler for the current state.

```
while True:
    if state == SET:
        handle_set_state()
    elif state == RUN:
        handle_run_state()
    elif state == ALARM:
        handle_alarm_state()
```

Switching Between States

Let's now expand the program so that when we press button B, we toggle back and forth between the SET and RUN states. The handlers for SET and RUN are listed in the next piece of code—you can run this example now using the program ch06_timer_3.py.

```
def handle_set_state():
    global state
    display.show("S")
    if button_b.was_pressed():
        state = RUN
```

The first thing to note is that the first line of the handler is:

```
    global state
```

By default, Python will allow you to read values from a global variable from within a function, but if you try to change the value of a variable inside a function, Python creates a local variable with the same name as the global variable unless you use the global command. Most programming languages do not do this; they allow unfettered access to global variables. So if you have come from another language and your Python program appears to be defying all logic, check that you are not accidentally "shadowing" a global variable.

The next line displays the letter "S" just so that we can see that we are in the SET state. If button B has been pressed (since last time we checked), the global variable state is set to RUN. The handler for the RUN state is pretty much a mirror image of the one for SET.

```
def handle_run_state():
    global state
    display.show("R")
    if button_b.was_pressed():
        state = SET
```

Adding Code to the SET State

Let's add a bit more to handle_set_state so that we can adjust the minutes. We are now reintroducing code that we originally wrote back in Chapter 4. You can find the revised program in ch06_timer_04.py. Here is the updated handle_set_state:

```
def handle_set_state():
    global state, mins
    if button_a.was_pressed():
        mins += 1
        if mins > 10:
            mins = 1
        display_mins(mins)
    if button_b.was_pressed():
        state = RUN
```

We have added mins to the list of global variables used because mins will be changed when you press button A. Whenever mins is changed, the function display_mins is called to update the display of minutes.

```
def display_mins(m):
    message = str(m)
    if m == 1:
        message += " min"
    else:
        message += " mins"
    display.scroll(message, wait=False, loop=True)
```

This is the same code that you met in Chapter 4, except that the optional parameters wait and loop have been added to scroll so that the message will repeatedly display the number of minutes set.

Try running ch06_timer_04.py, and notice how you can still switch between the modes using button B but can only change the minutes when you press button A.

Adding Code for the RUN State

As it stands, the RUN state simply displays an R to show you that you are in the RUN state and allows you to return to the SET state using button B. Referring to Figure 6-1, we also need to add the following to the handler for this state:

- Every second, decrease the count time (for which we will need a new `display_time`) function.

- Check for `mins` and `secs` to have reached zero, and then sound the buzzer and go to the `ALARM` state.

You can try out the revised version in the file `ch06_timer_5.py`. Before writing the handler itself, you are going to need a new function (`display_time`) that displays both the minutes and the seconds left in the countdown with a colon between them.

```
def display_time(m, s):
    message = str(m)
    if s < 10:
        message += ":0" + str(s)
    else:
        message += ":" + str(s)
    display.scroll(message, wait=False)
```

The message to display is always going to start with the number of minutes followed by a colon, but if the number of seconds is a single digit (<10), then we need a leading 0 before the seconds.

The `handle_run_state` function is by far the most complex in this program. Here is the revised function:

```
def handle_run_state():
    global state, mins, secs, last_tick_time
    if button_b.was_pressed():
        state = SET
        display_mins(mins)
    time_now = running_time()
    if time_now > last_tick_time + 5000:
        last_tick_time = time_now
        secs -= 5
        if secs < 0:
            secs = 55
            mins -= 1
        display_time(mins, secs)
    if mins == 0 and secs == 0:
        state = ALARM
        display.show(Image.HAPPY)
```

The first thing to notice is that there are two new global variables: secs, which records the current seconds count, and last_tick_time, which is used later in the function to determine whether it's time to "tick" the clock, reducing the time and updating the display. Both of these new variables are declared near the top of the program.

After the check for button A being pressed, there is a new section of code that calls a built-in function called running_time that returns the number of milliseconds since the micro:bit was last reset. This value is assigned to the variable time_now. If this time_now is greater than last_tick_time by 5 seconds (5,000 milliseconds), then first of all last_tick_time is set to time_now (ready for the next 5-second period to elapse), and then the number of seconds is reduced by 5. The timer counts down in steps of 5 seconds rather than the 1 second suggested by Figure 6-1 because it takes longer than a second for the display to scroll the time remaining.

The final part of handle_run_state checks to see if mins and secs are both 0, and if they are, it sets the state to ALARM and displays the HAPPY image on the display. In Chapter 10, we will modify this example again to make an alarm sound when the countdown is complete.

Adding Code for the ALARM State

At the moment when the countdown is complete, the program will go to the ALARM state and just sit there until the micro:bit is reset. The handle_alarm_state needs to be modified as shown in the next piece of code so that when button B is pressed, the program jumps all the way back to the SET state.

```
def handle_alarm_state():
    global state, mins
    if button_b.was_pressed():
        state = SET
        mins = 1
        display_mins(mins)
```

Load the timer example thus far from the file ch06_timer_06.py, and try it out.

Debugging

Programs rarely work perfectly first time. There is usually some *debugging* to be done to fix problems with the program not working correctly. The trouble is that it's often hard to see what's going wrong. If your bug actually causes an error message, then the REPL can be really useful in finding out what went wrong.

For example, load the program `ch06_debug.py`, but before you flash it onto your micro:bit, open the REPL.

```python
from microbit import *

x = 0
l = ['a', 'b', 'c']

while True:
    if button_a.was_pressed():
        display.show(l[x])
        x += 1
```

This is a program destined to fail because if you keep pressing button A, when it gets to 3, there will be an "index out of range" exception. Try pressing button A to cause the error. When the error occurs, the error message will be displayed in the REPL and will also scroll itself across the micro:bit's display agonizingly slowly. Of more use, though, when it finishes its scrolling, the program exits, and you are left at the REPL command line (Figure 6-2).

Figure 6-2 *Debugging with the REPL.*

From here you can run MicroPython commands, including checking on the values of variables. Type 1 to see the List and x to see the value of the index position. It's easy to see why the program failed.

More Python

In a "getting started with" book such as this, it is not possible to cover everything about Python. However, there are a few more things about Python that I think it would be useful to mention.

Formatting

If you are putting together a complicated message to be displayed either using `display.scroll` or `print` to print something in the REPL, you can append strings together using the plus sign to build up your message. You'll need to remember to turn any numbers into strings first using `str`. For example:

```
>>> name = "Simon"
>>> number = 1234
>>> print("Name: " + name + ", number: " + str(number))
Name: Simon, number: 1234
```

An alternative way of doing this is to use the built-in `format` method on string. This method assumes that the string is a *format string* made up of plain text (which will be left unchanged) and placeholders, in the form of pairs of curly braces, for values that will be substituted into the string. The preceding example could also be written as:

```
>>> name = "Simon"
>>> number = 12345
>>> print("Name: {}, number: {}".format(name, number))
Name: Simon, number: 12345
```

There is actually a whole load more of string formatting options. If you find yourself needing to do something fancy, take a look at the Python 3 documentation on formatting strings at https://docs.python.org/3/library/string.html#format -string-syntax.

Exception Handling

Exceptions are things that go wrong unexpectedly in a program. In Chapter 5, you discovered that if you try to access an element of a List that doesn't exist, you get an error message. You can deliberately cause this by typing the following in the REPL:

```
>>> l = [0, 1, 2]
>>> l[3]
Traceback (most recent call last):
File "<stdin>", line 1, in <module>
IndexError: list index out of range
```

What happens when you do something like this, which is not allowed, is that MicroPython is said to *raise an exception.* That is, the normal flow of the program is interrupted to flag the error. The program then stops.

Load the program ch06_exceptions.py and upload it onto your micro:bit with the REPL open so that you can see the messages it produces. Here is the program:

```
l = ['a', 'b', 'c']

try:
    print("1. Inside try")
    print("2. About to do something bad")
    l[3]
    print("3. After doing something bad")
except:
    print("4. Caught the exception")
print("5. And exiting")
```

And here is the resulting output:

```
1. Inside try
2. About to do something bad
4. Caught the exception
5. And exiting
```

Using try and except in this way allows you to catch the exception and prevent it from crashing the program. As you can see, output message 3 does not appear because as soon as the out-of-bounds exception happens, the program jumps away to the except block. After the except block has been run, the program continues as normal, with the last line of code to print message 5.

There is more to exceptions than I have covered here. You can, for instance, get hold of the error message in the `except` block and also only respond to certain types of exceptions and even deliberately raise your own exceptions. If you find yourself wanting to do something fancy with exceptions, take a look at https://python.readthedocs.io/en/latest/c-api/exceptions.html.

Generally, it's better to design your program so that exceptions do not get thrown, but sometimes this is outside the control of the program, as you will see in the next section.

File System

MicroPython can read from and write to files of data stored in a micro:bit's internal storage. You can create files from your program, write data into them, and then read them.

You can see the files on your micro:bit by clicking on the File button in Mu (Figure 6-3). The left side of the File area shows the files on your micro:bit (data files, not program code), and the right shows the files in the mu_code directory of Mu on your computer.

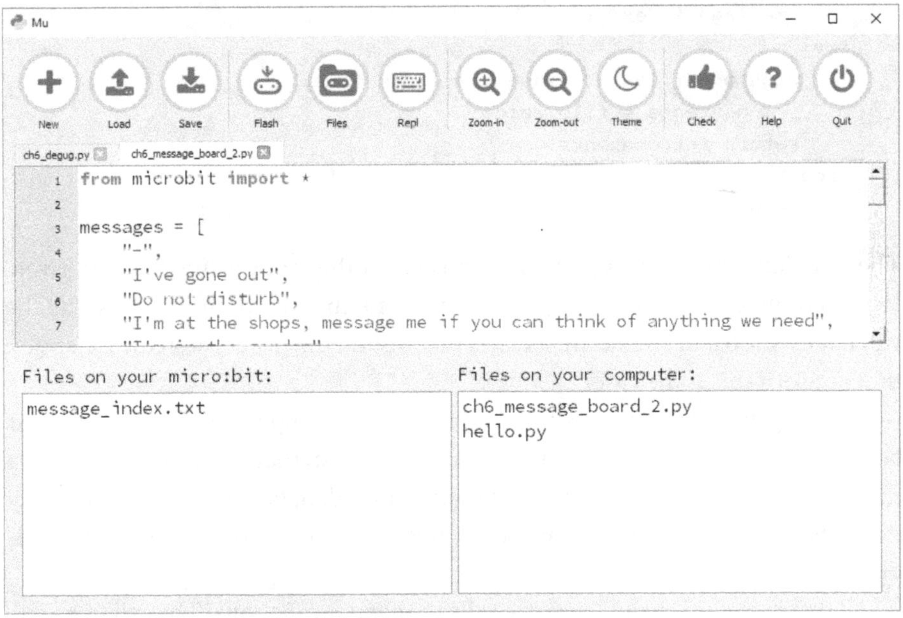

Figure 6-3 *The micro:bit file system.*

Files are a useful way of storing data that you don't want to be lost every time you restart your micro:bit. For example, let's modify the message board project of Chapter 5 so that it stores the index position of the message in a file. You can find this program in ch06_message_board_2.py.

The key changes to this project are the addition of two new functions: save_message_index and load_message_index. The first of these looks like this:

```python
def save_message_index():
    with open('message_index.txt', 'w') as file:
        file.write(str(message_index))
```

The with/open/as mechanism in Python is a neat way of ensuring that having opened a file to read, it is always closed so that other programs can use the file. The open part takes the name of the file to be written as one parameter and the mode to open it as its second (w for write); it then provides a handle to the file, in this case, the handle (also a variable) called just file. You can then write data to the file within the with/open/as block of code, and when you leave the block (stop indenting), the file is automatically saved and closed.

In this case, inside the block, the message_index is first converted into a string and then saved. Here is the counterpart for reading the file:

```python
def load_message_index():
    try:
        with open('message_index.txt') as file:
            contents = file.read()
        return int(contents)
    except:
        return 0
```

The function load_message_index will return the index value that was saved. However, there is a problem here. The first time that the program runs, the file will not exist, so a try/except block is put around the file to catch the exception that occurs if this is the case.

When opening a file to read it, you do not need to supply the second parameter because read mode is the default. If the file does exist, then the value is read from the file, converted into an integer, and then returned. If, however, the file does not exist, the exception is caught, and the value 0 returned, so the first message is displayed.

All that remains is to add in some calls to the new functions. The call to read_message_index only needs to occur once during startup, just before the message is displayed. However, the call to `write_message_index` needs to take place every time that the message index is changed. This all happens toward the end of the program. This is listed in the next piece of code with the new lines highlighted.

```
message_index = load_message_index()
show_current_message()

while True:
    if button_a.was_pressed():
        message_index += 1
        if message_index >= len(messages):
            message_index = 0
        save_message_index()
        show_current_message()
```

Try running the program, changing the selected message, and then unplugging your micro:bit. When you plug the micro:bit back in, your message choice will be displayed automatically. The file system on the micro:bit does not allow you to create directories. It's just a list of files. It also gets erased every time a new program is flashed onto the micro:bit.

Summary

In this chapter, we have delved a bit deeper into Python but somehow still managed to avoid one of the key features of this language, that is, *object orientation*. In Chapter 7, we will look at *classes* and learn a bit more about the built-in classes as well as how to make your own.

7

Modules and Classes

Modules are collections of code that are made available for you to use. The `microbit` module that you import at the top of pretty much all your programs is one such module. A module may just contain lots of functions and variables, but sometimes the module will contain classes. Classes provide a further level of structure. In this chapter, you will learn how all of this works and how to go about making use of modules and classes as well as writing your own.

Built-in Modules

When you start one of your programs with this line:

```
from microbit import *
```

you are telling MicroPython to use a module called `microbit` and import everything (*) from it. The `microbit` module is made up of classes that contain the methods you use to control your micro:bit.

The most important module for the micro:bit that we have been using throughout this book is, not surprisingly, `microbit`. This module contains all the code that relates to using the hardware of the micro:bit through MicroPython. The module is what makes it easy for us to scroll messages across the display or tell whether a button has been pressed or not. This is why most of your programs start with the line:

```
from microbit import *
```

In Chapter 8, you will use the module random, and in Chapter 10, you will use a couple of other built-in modules called speech and music, both of which have to be imported at the start of your programs if you need to use them.

You can think of importing a module in Python as taking the contents of a separate file of Python code (the module) and pasting it at the top of your program. Separating out code that could be useful into a separate file has the great benefit that the module can be reused in lots of your programs without having to copy and paste code every time. In reality, though, the importing mechanism is more subtle than this.

Importing Modules

A module is just another file full of Python code, and if we just want everything in that code to be included in our program, we can do that, but we run the risk that it will use variable or function names that are the same as ones that we have used. This would cause errors. Because we can't see the code for the microbit module, we are running a risk by making everything in the module visible.

Probably the most convenient way of knowing what is in the microbit (or any other) module is to find its documentation. This will tell us the things that are part of the module. By the way, you can find the documentation for the microbit module at https://microbit-micropython.readthedocs.io/en/latest/microbit.html.

Another useful way of finding what's in a module is to use Python's built-in dir function. You can try this in the REPL now.

```
>>> import microbit
>>> dir(microbit)
['__name__', 'Image', 'display', 'button_a', 'button_b',
'accelerometer', 'compass', 'i2c', 'uart', 'spi', 'reset', 'sleep',
'running_time', 'panic', 'temperature', 'pin0', 'pin1', 'pin2',
'pin3', 'pin4', 'pin5', 'pin6', 'pin7', 'pin8', 'pin9', 'pin10',
'pin11', 'pin12', 'pin13', 'pin14', 'pin15', 'pin16', 'pin19',
'pin20']
```

This shows us all the names that the module uses. So, if you import everything from the module using this line at the top of your program:

```
from microbit import *
```

you cannot make a global variable, function, or class (more on classes later) with the same name as any of the names on the list.

You can import selectively from a module if you want. In the preceding REPL example, the line:

```
import microbit
```

imports the `microbit` module, so it is available to use, but it does not actually import the names that are used in your program. This means that you can still use `scroll` in your program, but instead of writing just:

```
display.scroll("Hello World")
```

you have to put the module name at the front of the line like this:

```
microbit.display.scroll("Hello World")
```

If you know that you want to use `display` and `button_a` but nothing else from the `microbit` module, then you could specify this in the import like this:

```
from microbit import display, button_a
display.scroll("Hello")
```

In most cases, however, it's easiest just to import everything, especially in a package like `microbit`, where you are likely to use the majority of the module's names in your program.

Classes and Instances

When you use `button_a` and `button_b` in your programs, they are both *instances* of the class `MicroBitButton`. Note that these button instances use the same naming convention as variables (lowercase letters with undersores between the words); classes use an uppercase initial letter and start each new word with an uppercase letter.

The reason that `button_a` and `button_b` look like variables is because that's exactly what they are. They are variables defined in the `microbit` module. These two instances of `MicroBitButton` come ready to use and are the only two instances of `MicroBitButton`. After all, the micro:bit only has two buttons.

Similarly, `display` is the one and only instance of `MicroBitDisplay`. You can find out what class something is using the `type` function in the REPL. Try entering the following code. As you can see, pretty much everything in Python is an instance of some kind of class.

```
>>> type(button_a)
<class 'MicroBitButton'>
>>> type(display)
<class 'MicroBitDisplay'>
>>> type("abc")
<class 'str'>
>>> type(123)
<class 'int'>
>>> type(True)
<class 'bool'>
```

So what exactly is a class? A *class* contains a mixture of:

- **Attributes** (variables that belong to the class and should only be accessed via the class)

- **Methods** (functions that belong to the class and should only be accessed via the class)

For example, if you were to look at the `MicroBitButton` class source code, you would find no attributes, but you would find the method `was_pressed`, among others.

Inheritance

Classes are arranged in a hierarchy, and one class can "inherit" methods and variables from another class. There is a good example of this in the built-in classes `MicroBitDigitalPin` and `MicroBitTouchPin`. Both of these classes are used to represent the connection pins on the edge connector (you will find much more on the edge connector in Chapter 10). Just like `display`, `button_a`, and `button_b`, the pins on the micro:bit's edge connector are also accessed by the use of variables defined in the `microbit` package. Try running the following code in the REPL:

```
>>> type(pin9)
<class 'MicroBitDigitalPin'>
>>> type(pin2)
<class 'MicroBitTouchPin'>
```

This tells us that `pin2` (the center hole on the edge connector) is an instance of `MicroBitTouchPin`, whereas `pin9` (see Figure 10-17) is an instance of `MicroBitDigitalPin`. They are different classes because they have different capabilities. `pin2` can be used as a touch pin, whereas `pin9` can only be used as a normal digital pin.

To confirm the methods that each pin has access to, you can use the `dir` function as follows:

```
>>> dir(pin9)
['write_digital', 'read_digital', 'write_analog', 'set_analog_period',
'set_analog_period_microseconds', 'get_analog_period_microseconds']
>>> dir(pin0)
['write_digital', 'read_digital', 'write_analog', 'read_analog',
'set_analog_period', 'set_analog_period_microseconds', 'is_touched']
```

The two classes have most of the same methods, but the class that `pin0` is an instance of has two extra methods, `read_analog` and `is_touched`. It would be wasteful if both `MicroBitTouchPin` and `MicroBitDigitalPin` implemented the methods that they had in common, and to avoid this repetition, a mechanism called *inheritance* is used. This allows one class (`MicroBitTouchPin`) to inherit the methods and variables from another class (`MicroBitDigitalPin`). Figure 7-1 shows how the classes further down the hierarchy inherit methods from their parents.

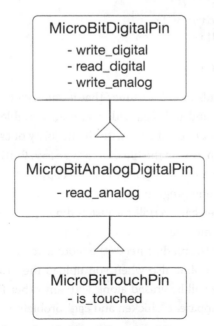

Figure 7-1 *The class hierarchy of pins on the micro:bit.*

Please note that I have shown how these classes are organized so that you can see how neat inheritance is. You don't ever actually need to create instances of any of these classes because all the instances you can ever use are already defined as pin0, pin1, and so on in the microbit module. Even though you don't need to use it, its nice to know that this has been implemented efficiently using inheritance.

Making Simple Modules

You may remember that in our timer example we had the problem of displaying either min or mins after the number of minutes depending on whether it was more than one minute or not. One (somewhat excessive) way to do this would be to define a function called pluralize that takes a string as a parameter, adds an s to the end of it, and then returns the result. This is a general-purpose function that could be of use in other programs, so, as an example, let's build it into a module and then see how we could make the timer example use it.

This function looks like this:

```
def pluralize(word):
    if word.endswith("s"):
        return word
    elif word.endswith("y"):
        return word[:-1] + "ies"
    else:
        return word + "s"
```

If the word already ends in s, it assumes that it's already plural; if it ends in y, it removes the last letter and adds ies. Otherwise, an s is added to the word. This is a long way from perfect but will cope with the majority of cases.

You can find the function in the file pluralize.py. Putting it in a separate file is enough to make it a module. We now need to copy it onto the micro:bit so that we can use it for a second program, which can be a modified version of the timer program. This is somewhat overkill for just adding an s to min, but it illustrates the process of using a module.

The program using the module needs the module to be on the micro:bit to run without giving an error, but also when you upload the program, reflashing the micro:bit, the module will be erased along with any other files on the micro:bit. The way around this apparent "chicken and egg" problem is to:

1. Flash the program that's going to use the module onto the micro:bit and allow it to fail with an error because the module is missing.

2. Copy the module file onto the micro:bit.

3. Press the Reset button so that when the program runs, it now succeeds because the module file is now on the micro:bit too.

Unfortunately, you will have to do this every time you modify the program using the module.

You can find a version of the timer project that is modified to use the `pluralize` module in `ch7_timer_using_module.py`. The first change to the timer project code is an extra import at the top.

```
from microbit import *
from pluralize import *
```

The other change is to the `display_mins` function so that it uses the module's function.

```
def display_mins(m):
    message = str(m)
    if m == 1:
        message += " min"
    else:
        message += pluralize(" min")
    display.scroll(message, wait=False, loop=True)
```

Upload the program to your micro:bit, and don't worry about the error message—that's just because the module's not there yet. Now click the File button on Mu (make sure that the REPL area is closed first), and drag the file `pluralize.py` from the Files on Your Computer area on the right to the Files on Your micro:bit" area to the left (Figure 7-2). After copying, the module file should appear in the left-hand area where you dragged it.

Note that I encountered a few problems using the File feature of Mu. If the file fails to copy, you may have to reset the micro:bit and/or quit and restart Mu. You can now press the Reset button on the micro:bit, and the timer program should function just as it did before, but now using the `pluralize` library.

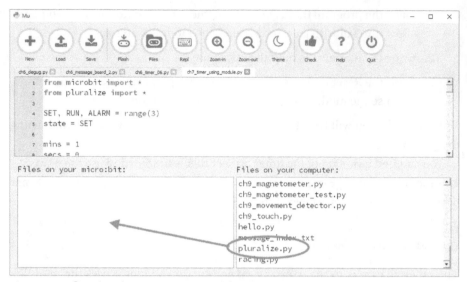

Figure 7-2 *Copying the* `pluralize` *module onto a micro:bit.*

Making a Module That Uses a Class

Although a module can be (and often is) a set of functions and/or variables, the `microbit` module clearly contains more than just variables and functions. We know this because we can write code that uses the module that looks like this:

```
display.scroll("Hello World")
```

So unless you can have function names with a dot in them (which you can't), there must be something more going on. The dot indicates the calling of a special type of function that belongs to a class of which `display` is one instance.

To illustrate how classes can be used in a module, let's design a class that implements a very simple menu system on the micro:bit. The idea is that you use button A to cycle round the menu options, each one being scrolled across the display. Then, when the menu option you want is being displayed, pressing button B will make the selection.

Here is an example of how you might use the module to ask for an image to display. You can find the program in `ch07_menu_example.py`. But to run it on the micro:bit, you will also need to copy the module file menu.py onto your micro:bit.

```
from microbit import *
from menu import *

image_menu = Menu({"Happy" : Image.HAPPY, "Sad" : Image.SAD,
                   "Angry" : Image.ANGRY})

while True:
    answer = image_menu.ask()
    display.show(answer)
    sleep(5000)
```

The menu is supplied with a dictionary at the time it is created, and the keys to the dictionary are the names of the images (Happy, Sad, and Angry), and the values are the three corresponding images. To get the micro:bit to display the menu, the method ask is called, and when a selection has been made from the menu, the ask function will return the image to be displayed. In the following sections, we will gradually build up the menu module.

Module and Class Definition

This module is going to contain the class Menu and, just for good measure, will also define one instance of the class called Yes_No_Menu that makes a menu with two options: "Yes", which returns True when selected, and "No", which returns False. Because the menu module makes use of the microbit module, it must import it.

To define a class, you use the word class followed by the name of the class and a colon. All the methods within the class are then indented.

```
from microbit import *
class Menu:
# class definition will go here
yes_no_menu = Menu({"Yes" : True, "No" : False})
```

Methods

Methods have the same structure as functions, except that every method must have a first parameter called self. This is necessary even if the method doesn't have any other parameters. The word self denotes the object itself and is a way for methods to access other methods and variables that belong to the class.

Most classes will need an initializer method called __init__. That's init with two underscore characters on each side of it. This method will be run every time a new instance of the class is created and is where you put any code that needs to be run to set up the new instance.

```
def __init__(self, choices):
    self.choices = choices
    self.selection = 0
    self.num_choices = len(choices)
```

In this case, the __init__ method takes the second parameter called choices and makes a new *instance variable* also called choices. This second choices is not a parameter like the first choices but a bit like a global variable, except rather than being global to the whole file, choices actually belongs to an instance of the class. A second instance variable called selection is also created and is assigned the value 0 to indicate that by default the first item in the menu is to be selected. The final instance variable, called num_choices is set to the length of the choices Dictionary.

The next method __watch_for_selection has a name that, like __init, is also prefixed by two underscores, indicating that the method is *private*. Being private means that the method should not be called from outside the class; it is if you like a helper method to be used by the ask method, but only the ask method is *public* (can be used from outside the class). Both methods are listed next:

```
def __watch_for_selection(self):
    if button_a.was_pressed():
        self.selection += 1
        if self.selection >= self.num_choices:
            self.selection = 0
        key = list(self.choices)[self.selection]
        display.scroll(key, loop=True, wait=False)


def ask(self):
    key = list(self.choices)[self.selection]
    display.scroll(key, loop=True, wait=False)
    while button_b.was_pressed() == False:
        self.__watch_for_selection()
    key = list(self.choices)[self.selection]
    return self.choices[key]
```

Let's start in reverse order, looking at `ask` first. This first finds the currently selected key from the Dictionary and displays it. Note the use of `self` before all references to the instance variables. Next, the `while` loop keeps calling `__watch_for_selection` until button B has been pressed, indicating that a final selection has been made. When this happens, the newly selected key is found, and the value associated with that key is returned.

The method `__watch_for_selection` checks to see if button A has been pressed, and if it has, it adds 1 to `self.selection`. It also checks whether the end of the menu options has been reached, and if it has, it sets the instance variable `self.selection` back to 0. Finally, it displays the changed menu selection.

Using the Module

If you want to try out the example, you will need to carry out the procedure described earlier in the section "Making Simple Modules." Alternatively, just to try out the code, you can forget that the module is a module and just paste the whole class definition at the top of your program.

Modules from the Community

One of the really nice things about most software communities is that people are willing to share the code they create so that other people can make use of it. Their Python modules are often referred to as *libraries*, and you will find a list of libraries and other micro:bit resources at https://github.com/carlosperate/awesome -microbit.

Most libraries that are contributed are concerned with connecting some external device, such as an ultrasonic rangefinder or a certain type of display. You will find more on using hardware in Chapter 10.

Summary

This chapter has explained the basics of modules as well as classes, instances, and methods. You have seen how the built-in `microbit` module makes use of these. In Chapter 8, we will concentrate on the micro:bit's LED display and the various ways that it can be used.

8

The LED Display

You have already used the LED display quite a lot. However, there are still quite a few features of this useful little matrix of LEDs to explore. You can use it to scroll text messages, display images, make animated displays, and even play simple games.

Controlling Individual LEDs

Let's start by experimenting with the REPL. Try the following commands, watching what happens to the LEDs between each step:

```
>>> from microbit import *
>>> display.clear()
>>> display.set_pixel(0, 0, 9)
>>> display.set_pixel(0, 0, 5)
>>> display.set_pixel(0, 0, 0)
```

display.clear() turns off all the LEDs in the display. The first set_pixel command sets the brightness of the LED at position 0, 0 to a brightness of 9 (maximum). The line after sets it to half brightness, and the final line turns the LED off.

The LEDs are accessed by their column (x position) and row (y position) numbers. The LED at the top left is LED 0, 0, the LED to its right is LED 1, 0, and so on. Figure 8-1 shows the coordinates of the LEDs.

Flash the program ch08_random_pixels.py onto your micro:bit. This produces a pleasing effect of random pixels turning on and off on the display.

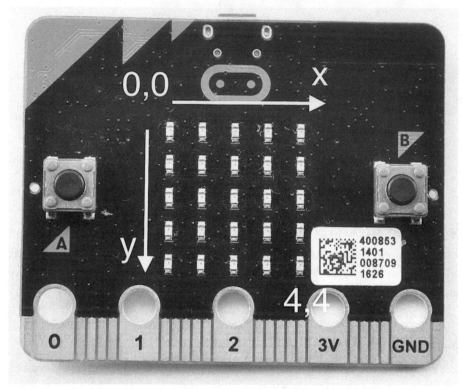

Figure 8-1 LED display coordinates.

```
from microbit import *
from random import randint

display.clear()

while True:
    x = randint(0, 4)
    y = randint(0, 4)
    brightness = randint(0, 1) * 9
    display.set_pixel(x, y, brightness)
    sleep(50)
```

The x and y coordinates to turn the LED on and off are set using randint (see the section "Numbers" in Chapter 3). We want the brightness to be either 0 or 9, so we first create a random number that is either 0 or 1 and then multiply it by 9. This works because $0 \times 9 = 0$ and $1 \times 9 = 9$.

Scrolling Text

At its simplest, to display a scrolling text message, you can just type the following in the REPL:

```
>>> display.scroll("hello world")
```

The message will scroll across the screen at a pace that is easy to read but a little on the slow side. You can control the scrolling speed using the optional parameter of `delay`, which sets the delay in milliseconds between each scrolling step. By default, this is set to 150 ms. Try changing it to 50 ms.

```
>>> display.scroll("hello world", delay=50)
```

Suddenly the message whizzes past much faster and is still just about readable. Experiment with a few different values of `delay`.

By default, the micro:bit cannot do anything else while the display is scrolling. The call to `scroll` is said to be *blocking* because it blocks anything else from happening on the micro:bit. This includes responding immediately to button presses, gestures, or anything else that you might want to be going on while a message is being displayed. The optional parameter `wait` can be set to `False` to force your program to continue with the next line of code without waiting for the scrolling message to finish.

The optional parameter `loop` will instruct the micro:bit to repeat the message until the message is replaced with a different one or `clear` is called. The `loop` and `wait` parameters are often used together, as they are in ch08_scroll_no_wait.py.

```python
from microbit import *

display.scroll("Press A to start")

while True:
    if button_a.was_pressed():
        display.scroll("Press B to stop", wait=False, loop=True)
    if button_b.was_pressed():
        display.clear()
```

Try out this example. After the initial normal blocking message, if you press button A, a message will start scrolling and repeat until you press button B.

Showing Text

The best way to display text of more than a single character is to use scroll. However, if you just want to display a single letter or number, then use show. Try this out in the REPL:

```
>>> display.show("A")
```

This can also be useful for displaying numbers, as in the example ch08_counting.py. However, you will need to use str to convert the digit into a string before using it with show.

```
from microbit import *

for x in range(0, 10):
    display.show(str(x))
    sleep(1000)
```

You can actually achieve the same effect as ch08_counting.py as follows. Try it in the REPL.

```
>>> display.show("0123456789", delay=1000)
```

Here each letter of the string of digits is displayed in turn. The optional delay parameter sets a 1-second delay (1,000 milliseconds) between each digit.

Showing an Image

There are a whole load of images bundled with MicroPython. To display an image, use show like this:

```
display.show(Image.HEART)
```

Figure 8-2 shows the HEART image.

You can find a full list of built-in images at https://microbit-micropython .readthedocs.io/en/latest/image.html#attributes, but some of the more useful ones include

- Image.HAPPY—smiley face
- Image.SAD—sad face

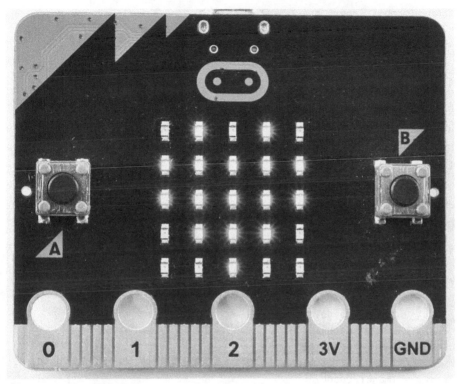

Figure 8-2 *The HEART image.*

- `Image.YES`—checkmark
- `Image.NO`—cross

Some of the images (`Image.RABBIT`, for example) require a fair amount of imagination.

Animation

Animations are basically sequences of images shown in quick succession. Load the example `ch08_clock.py`, and run it. You will see the display indicate 12 positions around the face of a clock. This is fairly crude, but the best you can do with 25 pixels.

```
from microbit import *

while True:
    display.show(Image.ALL_CLOCKS, delay=1000)
```

And yes, just these few lines of code are needed to do this. This is helped by the fact that `Image.ALL_CLOCKS` returns not just a single image but a List of 12 images. The `show` method, when provided with a List of images, will display one image after the other. The optional parameter `delay=1000` slows it down to a 1-second tick.

To make your own animation, you just need to create a List of images. Load the example `ch08_animate.py`. This will animate a square expanding and then contracting.

```
from microbit import *

frame1 = Image("00000:"
               "00000:"
               "00900:"
               "00000:"
               "00000:")

frame2 = Image("00000:"
               "09990:"
               "09090:"
               "09990:"
               "00000:")

frame3 = Image("99999:"
               "90009:"
               "90009:"
               "90009:"
               "99999:")

frames = [frame1, frame2, frame3, frame2]

while True:
    display.show(frames, delay=200)
```

Each of the frames is created as an image, and then the `frames` List combines them all in the correct order. Note that frame 2 is used twice in the List.

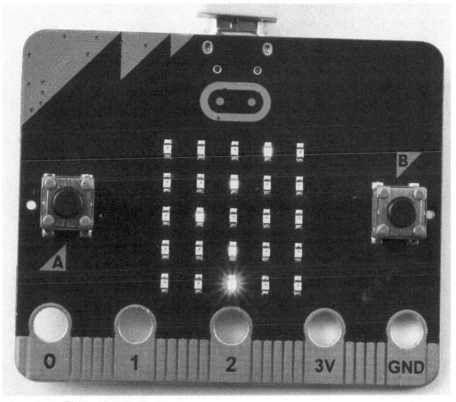

Figure 8-3 *The racing game.*

Racing Game

Let's end this chapter with a simple game (Figure 8-3) that is surprisingly addic-tive. Buttons A and B are used to steer a "car" (bright dot) left and right at the bottom of the screen while obstacles (dimmer dots) appear at the top of the screen and scroll toward the car. The aim is to keep going as long as possible before hit-ting something. To make it harder, the obstacles move faster and faster as time goes on. You will find the program in `ch08_racing.py`.

```
from microbit import *
from random import *

track = Image("00000:"
              "00000:"
```

```
                    "00000:"
                    "00000:"
                    "00000")

car_position = 2
delay = 500
score = 0

while True:
    track.set_pixel(randint(0, 4), 0, 5)
    if button_b.was_pressed() and car_position < 4:
        car_position += 1
    if button_a.was_pressed() and car_position > 0:
        car_position -= 1
    if track.get_pixel(car_position, 4) > 0:
        display.scroll("GAME OVER!" + str(score))
        break
    track.set_pixel(car_position, 4, 9)
    display.show(track)
    track = track.shift_down(1)
    sleep(delay)
    delay -= 1
    score += 1
```

The variable track contains a blank image. It is this image that will be displayed on the screen. The reason that the screen is not modified directly is that we want to be able to make use of the shift_down method to move the approaching obstacles down the screen.

The x position of the car is kept in the variable car_position. The variable delay holds the delay period in milliseconds between successive updates of the game. This starts at 500 ms and decreases by 1 ms for each frame of the game. Similarly, the variable score starts at 0 and then increases by 1 until you crash into an obstacle.

The first thing that happens inside the while loop is that an obstacle is introduced on the top line at a random position. Next, both buttons are checked, and car_position is moved left or right in response.

To check whether you have hit an obstacle, the pixel at the car's position is checked, and if the pixel is anything other than off, a collision has happened and the message GAME OVER! followed by the score is scrolled across the display. To play again, press the Reset button.

Summary

The microbit library makes is super easy to use the LED display, and it's surprising what you can do with it despite its very low resolution. In Chapter 9, you will learn more about the various sensors on the micro:bit.

9

micro:bit Sensors

One way that the micro:bit scores over rivals such as the Arduino is that it has a number of built-in sensors that you can use without having to attach any external electronics. It has an accelerometer that you can use to detect movement gestures and a magnetic compass as well as the ability to detect touch. In this chapter, you will explore these various sensors.

Buttons Revisited

Before we look at the real sensors, let's look in a little more detail at the built-in A and B buttons. So far we have used the was_pressed method to detect button presses, but there are some other methods:

- is_pressed() returns True if the button is actually pressed at the time the method is called.

- was_pressed() returns True if the button has been pressed at some time between when you last called this method and the time you are calling it now. Using this method means that you can't "miss" button presses because the program was doing something else when the button was being pressed.

- get_presses() returns the total number of presses since the last time you asked.

Gestures

Gestures are the easiest way to use the micro:bit's accelerometer. The most common use of the accelerometer is to check for the orientation of the micro:bit and sudden events such as the micro:bit being dropped, both of which can be done using gestures.

We first met gestures back in Chapter 5 in ch05_gesture_detector.py, which displays the orientation of the micro:bit as you tip it one way or another. Other gestures detect the levels of shock (maximum acceleration exceeding 3*g* [three times gravity], 6*g*, and 8*g*) as well as freefall. Freefall is difficult to test without access to space travel or a desire to drop your micro:bit onto the floor (both risky things to do). Similarly, the acceleration gestures of 3*g*, 6*g*, and 8*g* are also quite hard to test without giving your micro:bit a good bashing.

Perhaps the most useful of these other gestures is *shake*. As the name suggests, this just detects that the micro:bit has been shaken. You can use it to make a very simple die that displays a random number between 1 and 6 each time you shake the micro:bit (Figure 9-1). You can find this in ch09_dice.py.

```
from microbit import *
from random import randint

while True:
    gesture = accelerometer.current_gesture()
    if gesture == "shake":
        display.show(str(randint(1, 6)))
```

Raw Accelerometer Data

In addition to using the micro:bit's gesture detection, you can also gain access to the raw data from the accelerometer if what you want to do can't be done using gestures. To get a feel for the type of data you get back from the accelerometer, run the program ch09_accl_test.py with the REPL open so that you can see the output.

```
from microbit import *
while True:
    x, y, z = accelerometer.get_values()
    print('x={}, y={}, z={}'.format(x, y, z))
    sleep(5000)
```

Figure 9-1 *A simple die.*

You should see output like this if your micro:bit is lying more or less flat:

```
x=16, y=48, z=-976
x=0, y=48, z=-976
x=-32, y=48, z=-992
x=32, y=80, z=-976
x=16, y=64, z=-976
```

Try moving the micro:bit around to see how the readings change. Figure 9-2 shows how the accelerometer works.

The accelerometer actually measures the forces on a tiny weight contained in the accelerometer chip. The constant (and biggest) force acting on the

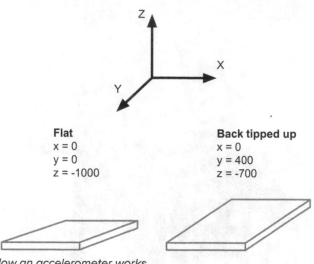

Figure 9-2 *How an accelerometer works.*

accelerometer is the force of gravity. With the micro:bit level, this force is in the z
dimension. This is vertical. The y direction is from front to back of the micro:bit,
and the x direction is from left to right. Note that the x and y are the same as for
the display. When you tip the micro:bit (and hence its accelerometer chip) a little,
some of the force of gravity starts to act on the other dimensions, changing the x
and y readings.

The program ch09_movement_detector.py monitors the acceleration on
just the x axis, and if it changes by more than a certain threshold, a message is
displayed. Increasing the value of THRESHOLD will make the detector less sensitive. See if you can pick up your micro:bit without triggering the alarm.

```
from microbit import *

THRESHOLD = 100

old_x = accelerometer.get_x()

while True:
    x = accelerometer.get_x()
    if abs(x - old_x) > THRESHOLD:
        display.scroll("ALARM!")
    old_x = x
```

Although I picked the value of acceleration in the x dimension, any of the other dimensions would work equally well.

Note the use of the abs built-in function when checking whether the change in reading has exceeded the threshold. The abs function returns the *absolute value* of a number, which is just a fancy way of saying that if it's negative, it removes the minus sign.

After reading Chapter 10, you might like to return to this project and add a buzzer or speaker that sounds when the micro:bit is moved.

Magnetometer

In addition to an accelerometer, the micro:bit also has a built-in three-axis magnetometer (or digital compass) chip. This senses the strength of a magnetic field in each of the three x, y, and z directions. This can be used as a compass, and at its simplest, you can just use the method compass.heading, which returns the bearing as an angle between 0 and 360 degrees, where both 0 and 360 degrees are magnetic north. If your phone has a compass app and you've tried using it, then you'll have an idea of how useful it is as a compass. Generally speaking, the results are not great with the micro:bit, and like a phone app, you will need to calibrate the compass using the method compass.calibrate. The example ch09_compass.py shows a ^ on the display when your micro:bit is facing "northish" and a V when it is facing more or less south.

```
from microbit import *

angle = 10

while True:
    if button_a.is_pressed() and button_b.is_pressed():
        compass.calibrate()
    heading = compass.heading()
    if heading >= (360 - angle) or heading <= angle:
        display.show("^")
        sleep(100)
    elif heading >= (180 - angle) and heading <= (180 + angle):
        display.show("V")
        sleep(100)
    else:
        display.clear()
```

The compass knows whether it has been calibrated, and if it hasn't—or a new program has been flashed onto the micro:bit—it starts a calibration routine that asks you to tilt the micro:bit so as to produce a circle of dots around the outside of the display. Once this is done, the calibration will be remembered, but you can force a recalibration at any time by pressing both micro:bit buttons at the same time.

In the same way as the accelerometer has a high-level gesture interface and also low-level routines to read the raw data, the compass also has a low-level interface that allows you to read the strength of magnetic field in the x, y, and z directions. You can use this low-level interface to detect a magnet (for best results, use a "neo" neodymium magnet) and even to get an idea of how near or far the magnet is from the micro:bit (Figure 9-3).

Figure 9-3 *Detecting a magnet.*

You will find the program for this in `ch09_magnetometer.py`. Try moving a magnet near button B (the magnetometer chip is close to button B on the underside of the board), and you should see the number of bars change as you move the magnet closer and further away.

```python
from microbit import *
from math import sqrt

# Reduce to make more sensitive
MAGNET_MAX = 100000

baseline = compass.get_z()

def scale_inv_square(a, a_max, scaled_max=5.0):
    a_max_scaled = sqrt(abs(a_max))
    a_scaled = sqrt(abs(a))
    return int(a_scaled * scaled_max / a_max_scaled)

def bargraph(a):
    display.clear()
    for y in range(0, 5):
        if a > y:
            for x in range(0, 5):
                display.set_pixel(x, 4-y, 9)
while True:
    z = compass.get_z()
    num_bars = scale_inv_square(z - baseline, MAGNET_MAX)
    bargraph(num_bars)
    if button_a.was_pressed():
        baseline = compass.get_z()
```

The global variable MAGNET_MAX is all in uppercase letters to highlight it as a constant. This is a value that won't change when the program is running, but you might want to change it before running the program. It is used to scale the reading and control how many rows of LEDs will be lit. As the comment above it suggests, reduce this value to make the meter more sensitive.

The global variable baseline is used to hold the background field strength read before a magnet is placed near the micro:bit. It is initialized to the z-axis reading of the magnetometer.

The strength of magnetic field at a certain distance from the magnet follows what is called an *inverse square law*. This means that if you start out 2 inches from the magnetometer and take a reading x and then move to 4 inches away (the distance has doubled) and now take another reading (y), y will not be one-half of x but one-quarter of x. The function scale_inv_square takes three parameters: a reading from the magnetometer, the maximum expected reading, and the maximum value of result (scaled_max). The parameter scaled_max defaults to 5 to give five bars on the display. The function takes the square root of both the measurement and the maximum expected readings (after removing the signs using abs) and then returns the ratio scaled by scaled_max.

To display a number of rows between one and five for the reading, the function bargraph uses a for loop to count using y from 0 to 4, each time selecting a row. If the value supplied as the parameter a exceeds the current row number, then the whole row of pixels is lit by another for loop. Note that in the call to set_pixel, the y coordinate is 4-y to flip the display so that the bars start from the bottom of the display rather than the top.

The main loop takes the magnetometer reading on the z axis and then displays the difference between it and baseline. It then checks for a press of button A, and if there is one, it resets the value of baseline.

Processor Temperature

The micro:bit's microprocessor has a built-in temperature sensor. You can access this to measure the temperature, but it will give a reading that is somewhat higher than the ambient temperature. You can try out this feature in the REPL as follows:

```
>>> temperature()
29
```

The temperature is reported in degrees centigrade; to convert to Fahrenheit, multiply by 9/5 and then add 32.

Touch

You will learn more about using the edge connector in Chapter 10 when you start to attach external electronics to your micro:bit. However, you can use the points on the connector labeled 0, 1, and 2 as touch buttons (Figure 9-4).

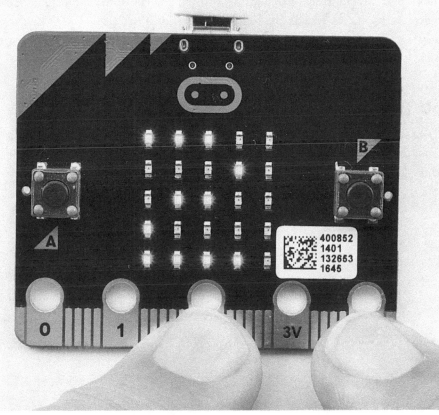

Figure 9-4 *Using the edge connector as touch buttons.*

Try out program ch09_touch.py to see how this works.

```
from microbit import *

while True:
    if pin0.is_touched():
        display.show("0")
    elif pin1.is_touched():
        display.show("1")
    elif pin2.is_touched():
        display.show("2")
```

These "pins" are not true touch buttons; you generally need to have a connection to ground (GND) as well as the touch pin, as shown in Figure 9-4. When you touch one of the pins, its number should appear on the display. If this doesn't work, it may be that your fingers are too dry.

Summary

In this chapter, you have seen some of the sensors that are built into the micro:bit's circuit board. In Chapter 10, you will take this further and start attaching some other electronic items to the micro:bit.

10

Connecting Electronics

You can go a long way with the micro:bit just using its built-in sensors and display, but at some point you will probably want to attach some external electronics—maybe to produce sound, power the micro:bit from a battery, connect other types of sensors, or turn something on and off.

Battery Power

Powering your micro:bit is great while you are getting your program working, but sometimes you will need to unplug it from your computer and take your project wireless, for which you will need to power your micro:bit with batteries.

USB Power Pack

One quick, easy (and rechargeable) way to power your micro:bit with batteries is to use a USB power pack such as the one shown in Figure 10-1.

A WORD OF WARNING: *The Microbit Foundation specifically advises against using USB battery packs (see www.microbit.co.uk/safety-advice). The foundation says that this is because it is possible to find USB power packs that supply more than the 5 volts (V) expected of them. The USB interface chip that regulates the USB input voltage from the nominal 5 V to 3.3 V (which all the other chips on the micro:bit require) is specified as allowing up to 6 V as its input voltage. In my experience, you would be extremely unlucky to exceed 6 V with a USB power pack because they have their own regulator chips to set their output voltage to a stable 5 V.*

Figure 10-1   *Powering a micro:bit from a USB power pack.*

It is more risky to use USB phone chargers, especially really cheap ones. These have been known to produce a higher voltage than 5 V to allow for a voltage drop in the wires to the phone to be charged. A micro:bit consumes much less current than a phone being charged and so does not suffer the same voltage drop in the wires. Thus it could receive over 6 V, possibly destroying the micro:bit.

Assuming that you are happy to take the risk of using a USB power pack, the micro:bit uses very little electricity (more on this later), so the smallest USB power pack you can find will be just fine. In fact, the larger, more expensive power packs tend to have an automatic shutoff feature, which means that they turn off if not enough current is being drawn from them, and sometimes this means that they don't detect the micro:bit and turn themselves off. The power pack in Figure 10-1 was bought for less than $5 and is charged from your computer or any USB power source (incidentally, I measured its output voltage as a very stable and perfectly acceptable 5.15 V).

Powering a micro:bit from a USB power pack actually provides a more stable voltage source to the micro:bit than using 3 V batteries, as described in the next subsection. This is so because the USB connection provides 5 V that is then regulated using a chip on the micro:bit to a precise 3.3 V supply, whereas when using a 3 V battery pack, the actual voltage will vary between 2.6 and 3.2 V depending on how fresh the batteries are.

3 V Battery Pack

The micro:bit has a special connector to which a battery pack can be attached. Figure 10-2 shows a 2 × AAA battery pack powering a micro:bit. Next to it is a second type of battery pack that very usefully incorporates a switch so that you can turn the micro:bit on and off without having to plug and unplug the connector. The battery socket has a slot on one side, so it can only be connected the correct way.

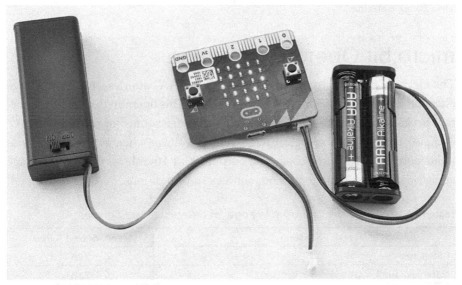

Figure 10-2   *3 V battery packs.*

Note that when you are powering a micro:bit from its 3 V connector, the power LED does not light, so it can be difficult to detect whether your batteries are okay unless you have something being shown on the display.

Battery Life

A micro:bit consumes about 19 milliamperes (mA) of current with none of the display LEDs lit. It then consumes a further 1 mA for each LED. So if you assume that half the LEDs will be lit on average, then the micro:bit will consume a total of around 30 mA. If you use the micro:bit's radio (see Chapter 11), you may use another 10 mA or so. Table 10-1 shows how long various power sources will power a micro:bit, assuming an average current consumption of 30 mA.

Table 10-1 *Battery Life When Powering a micro:bit*

Battery Type	Milliamp-hours (mAh)	Hours of Power
Cheapest of the cheap USB power pack	600	20
2 × AAA battery pack (low cost)	300	10
2 × AAA battery pack (alkaline, long life)	1,150	38
2 × AA battery pack (low cost)	600	20
2 × AA battery pack (alkaline, long life)	2,100	70

Note: For more information on the capacities of various types of battery, go to www.allaboutbatteries .com/Energy-tables.html.

micro:bit Operating Voltage

The means of powering your micro:bit—USB or AAA batteries—has quite a large effect on the *operating voltage* of your micro:bit. This operating voltage is important because it will be the actual voltage on the edge connector pin marked "3V" and also determines the logical HIGH output voltage when a general-purpose input/output (GPIO) pin is used as an output. Table 10-2 shows the actual voltage appearing on the 3 V pin for various sources and conditions.

Table 10-2 *The micro:bit Operating Voltage for Different Power Sources*

Source	Voltage on 3 V Pin
USB power	3.16
Battery power (2 × AAA fresh 3 V batteries)	2.85
Battery power (2 × AAA batteries part discharged to 2.64 V)	2.50

All the voltages measured on the 3 V pin are measured at about 0.14 V less than you might expect. The USB supply is regulated down to a precise 3.3 V by the USB interface chip, so you would expect the operating voltage to be 3.3 V, but instead it's 3.16 V. This is so because it takes 0.14 V for the circuit to allow automatic switching between battery and USB power.

GPIO Pins 0, 1, and 2

GPIO is short for "general-purpose input/output"—"general purpose" because such pins can operate in different ways:

- **Digital output**—turns something on or off

- **Analog output**—controls the amount of power going to something
- **Digital input**—detects an on/off signal such as a switch being pressed
- **Analog input**—measures a voltage

GPIO "pins" on the edge connector are clearly not very pin-like. They are gold-plated holes to which you can attach an alligator clip. The term *pin* comes from the fact that they are connected to pins on the micro:bit's microcontroller chip.

In this section, we will concentrate on the three easily accessible pins 0, 1, and 2. Later in this chapter, you will learn about the other pins available on the micro:bit's edge connector.

Alligator Clip Leads

Figure 10-3 show a lead terminated in alligator clips (also known as *crocodile clips*). These leads are really useful for attaching things to your micro:bit. Having a range of colors for your lead is also useful, and it's a good idea to make leads to 3 V red (for positive); leads to ground (GND) black, blue, or green; and other leads various other colors.

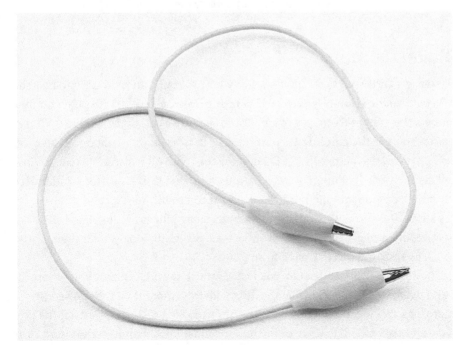

Figure 10-3 *A lead terminated in alligator clips.*

Alligator clips are so named because they resemble the jaws of an alligator. When you squeeze the back of the clip, the jaws at the front open and can be clipped onto the large hole connectors on the micro:bit edge connector to connect the micro:bit up to other devices. When you attach a clip to the edge connector of your micro:bit, it's best to do it vertically, as shown in Figure 10-4. Otherwise, the clip can accidentally connect the big pad on the edge connector to the smaller pads on either side.

Figure 10-4   *Connecting an alligator clip to a micro:bit.*

Digital Outputs

If you use a GPIO pin as a digital output, your program can set the pin to be HIGH (3 V, but see the section "micro:bit Operating Voltage") or LOW (0 V). If you then connect this output to one leg of a LED and the other end of the LED to 0 V (via a resistor to limit the current), then when the pin is HIGH (3 V), there will be a difference in voltage across the LED and resistor that will cause a current to flow, making the LED light up. (For information on buying LEDs, resistors, and indeed any other components and hardware for your micro:bit, see Appendix B.)

Figure 10-5 shows this arrangement built using alligator clips. The LED has a positive and a negative end; the positive lead is slightly longer than the negative lead. The LED will be controlled from pin0.

The reason that you should not just connect the LED directly between the GPIO pin and GND (0 V) is that the LED will draw more than the allowed (in the micro:bit's specification) 5 mA from the pin without a resistor to restrict the flow of current to 5 mA. The value of the resistor must be 330 ohms (Ω) or more. You probably could "get away" with attaching a LED directly, but doing so may damage

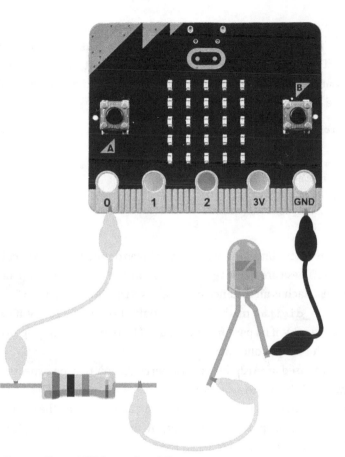

Figure 10-5 *Connecting a LED to a micro:bit.*

your micro:bit. It will not destroy it immediately, but the heating on the surface of the chip around the area responsible for the GPIO pins can shorten the life of the chip. In any case, using a resistor with a LED is a good habit to get into should you move on to using less forgiving boards.

The micro:bit's GPIO pins are actually quite robust, so even connecting a high digital output directly to GND only causes a current of 15 mA to flow. This is still three times what the micro:bit specifications recommend, but as long as you recognize your mistake and remove the short circuit, your micro:bit should survive.

When it comes to modern high-brightness LEDs, they are so efficient that even if you use a 1 kilo ohm (kΩ) resistor limiting the current to 1.7 mA, they will still be plenty bright enough and be well below the 5 mA maximum.

If you have a LED, a resistor, and some alligator clips, connect them together as shown in Figure 10-5. Flash the program ch10_blink.py onto your micro:bit, and the LED should blink once per second. If it doesn't, check that the LED is the right way around. Note that if it's the wrong way around, no harm will befall the LED or the micro:bit; the LED just won't work.

```
from microbit import *

while True:
    pin0.write_digital(True)
    sleep(500)
    pin0.write_digital(False)
    sleep(500)
```

To use the micro:bit's pins, you use variables defined in the microbit module for each pin. These are named pin0, pin1, and pin2 for the main alligator clip–friendly pins. Each is an instance of the class MicroBitTouchPin.

The write_digital method sets the output of the pin to 0 V if its parameter is False and to 3 V if its parameter is True. The two calls to sleep slow down the blinking of the LED to once per second.

If you have used an Arduino or Raspberry Pi, you may be expecting to have to set the mode of the pin to input or output before using it. This is not necessary with the micro:bit; if you try to use write_digital on a pin, the micro:bit assumes that you are using the pin as a digital output, and if you use read_digital with a particular pin, it sets that pin to be an input.

Because of the low current-handling capability of the GPIO pins (it's not recommended to draw more than 15 mA in total from these pins), digital outputs are usually used as signals to other electronic devices that then handle the larger current. If you have a larger current to switch, then you can use a relay board such as the MonkMakes Relay for micro:bit. Figure 10-6 shows one of these devices in action. A HIGH level at the IN connection makes an electronic switch integrated circuit (IC, called a *solid-state relay*) conduct so that the two connections marked OUT are connected together, completing a circuit to turn on a light bulb. The input signal only needs to supply about 3 mA, but the output can switch 2 A at up to 16 V. The input and output are not electrically connected.

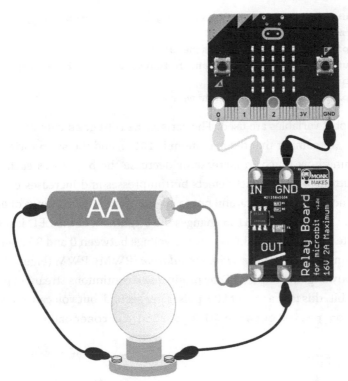

Figure 10-6 *Using a MonkMakes relay module.*

Analog Outputs

In addition to simply turning a LED on and off, you can also control the brightness of a LED by using the `write_analog` method instead of `write_digital`. Rather than simply accepting `True` or `False` for on and off, the `write_analog` method takes a value between 0 (off) and 1023 (full power).

Using the same arrangement of a LED attached to `pin0` as in Figure 10-5, run the program `ch10_brightness.py`. Pressing button B will increase the brightness, and pressing button A will decrease the brightness.

```
from microbit import *

brightness = 0
step = 100

while True:
```

```
if button_a.was_pressed() and brightness >= step:
    brightness -= step
    pin0.write_analog(brightness)
if button_b.was_pressed() and brightness <= (1023 - step):
    brightness += step
    pin0.write_analog(brightness)
```

Two global variables are used. The variable brightness contains the current brightness (remember that the maximum is 1023), and the step variable is used to determine how much to increase or decrease the brightness each time you press a button. The code just detects button presses and increases or decreases the brightness while being careful not to go out of the allowed range of 0 to 1023.

The way that write_analog changes the brightness of the LED is not (as you might quite reasonably think) to vary the voltage between 0 and 3 V, but rather it uses a technique called *pulse width modulation* (PWM). PWM (Figure 10-7) controls the power going to the LED by producing a continuous stream of pulses. On the micro:bit, this is at a rate of 50.3 pulses per second, but you can adjust it using set_analog_period or set_analog_period_microseconds.

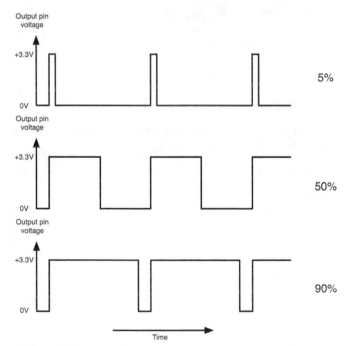

Figure 10-7 *Pulse width modulation.*

If the pulse is only HIGH for a very short time (say 5 percent of the time), then only a small amount of power is delivered to the LED, so it will be dim. The more time the pulse is HIGH, the more power is delivered to the LED, and the brighter it will be.

PWM also works for controlling the speed of motors, although motors use a lot more current than the 5 mA limit of a micro:bit's GPIO pin, so do not try to connect a motor directly. You could use a MonkMakes relay board to control a motor using PWM. This is possible because the MonkMakes Relay uses a solid-state relay with no moving parts. You should not use an old-fashioned electromechanical relay for PWM.

Digital Inputs

A push switch is a good example of a digital input to the micro:bit. Such a switch is digital in the sense that it can only be either on or off. It cannot be anywhere in between. The two switches built into the micro:bit are connected to GPIO pins (5 and 11) used as digital inputs.

You can test digital inputs by using a crocodile clip to connect pin0 to either 3 V or GND. Flash the program ch10_digital_input.py onto your micro:bit, and then try connecting the alligator lead from pin0 to 3 V (Figure 10-8A) and then from pin0 to GND (Figure 10-8B).

Be careful when connecting things to GPIO pins that are used as digital inputs because the voltage should not exceed 3.6 V, so connecting, say, 5 V to a digital input could destroy your micro:bit.

Analog Inputs

Whereas a digital input can only be on or off, analog inputs allow you to measure the voltage at a pin. The voltage must be between 0 V (GND) and the supply voltage (3V). Trying to measure higher voltages than this could destroy your micro:bit.

Flash the program ch10_voltmeter.py onto your micro:bit, and then try repeating the experiment of the preceding section, connecting pin0 first to GND and then to 3 V. You will see the voltage reading on the display change.

An AA or AAA battery has a voltage of around 1.5 V, so find yourself a battery, and use this project to measure the voltage of the battery. If like me you have a box full of batteries of indeterminate age and charge level, then you can use this as a battery tester. Connect the battery as shown in Figure 10-9.

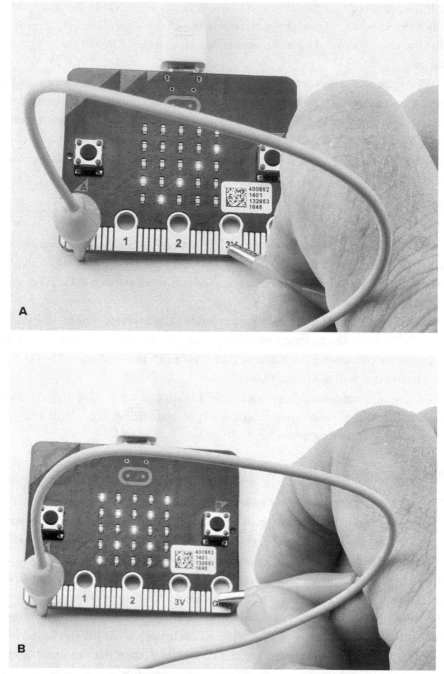

Figure 10-8 *Testing digital inputs on the micro:bit.*

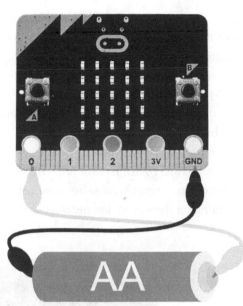

Figure 10-9 *Using a micro:bit as a battery tester.*

WARNING: *Be careful. If you connect the battery the wrong way around, you could destroy your micro:bit. The negative battery terminal must be connected to GND on the micro:bit.*

Here's the code for this project:

```
from microbit import *

while True:
    reading = pin0.read_analog() # 0 to 1023
    voltage = reading * 3.0 / 1023
    display.show("{:.1f}".format(voltage))
```

The function read_analog returns a number between 0 and 1023. A return number of 0 means 0 V, and if you are powering your micro:bit via USB, then a reading of 1023 would indicate 3.16 V (actually, the operating voltage). So if you get a reading of 511 (about half of 1023), then that would mean about 1.58 V.

To convert a reading value of between 0 and 1023 to a voltage of between 0 and 3 V, you need to first multiply the reading by 3.16 and then divide it by 1023. Finally, the voltage reading is formatted to one decimal place using the formatting string.

I mentioned earlier that if you are powering your micro:bit from its USB connector, then a reading of 1023 would indicate 3.16 V, the implication being that if you powered it from batteries, this would be something else. This is so because when you power your micro:bit powered from a battery pack, its supply voltage can be anything from 2.6 to 3.2 V depending on how fresh the batteries are. When your micro:bit is powered from batteries, a reading of 1023 indicates the operating voltage of the micro:bit.

Analog inputs are often used as a way of connecting sensors to a micro:bit. The MonkMakes sensor board (Figure 10-10) has sound, temperature, and light sensors, each of which produces an output voltage between zero and the micro:bit's operating voltage depending on the sound, temperature, or light level.

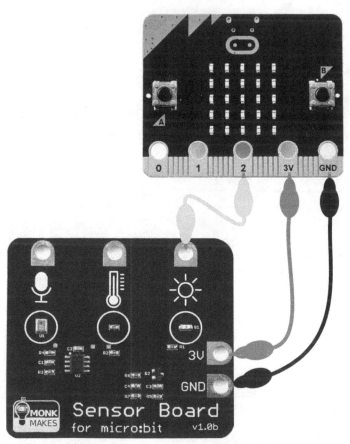

Figure 10-10   *The MonkMakes sensor board for micro:bit light sensing.*

If you have a photoresistor and fixed-value resistor, you can also make a light-sensing circuit if you connect them up as shown in Figure 10-11. Both this arrangement and the preceding one will work with the same program (ch10_voltmeter .py), with the voltage reading being an indication of the light level.

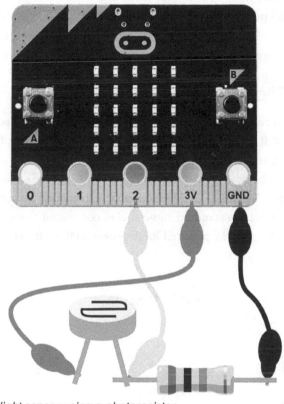

Figure 10-11 *A light sensor using a photoresistor.*

Power Out

When you power your micro:bit via USB, the regulator built into the USB interface chip on the micro:bit produces a steady 3.3 V for the micro:bit's various processors and sensors. The regulator is capable of supplying 120 mA before it gets too hot and turns itself off. This means that after the micro:bit has taken about 30 mA from the regulator, up to 90 mA will be available on the 3 V edge connector for your projects.

Sound Output

There are various ways of giving your micro:bit the ability to make a noise. In this section, we'll start by looking at the various connection options, and then we'll look at how you can write code to play tunes and even make your micro:bit talk. The micro:bit can produce an audio signal on one of its pins (by default pin0) by making the pin oscillate from LOW to HIGH hundreds or thousands of times a second. The faster the oscillation, the higher is the pitch of the sound produced.

Alligator Clip to Audio Socket

The GPIO output cannot supply enough current to be safely connected directly to a loudspeaker, but it can drive a small in-ear headphone. However, first you would need to connect one.

The easiest way to do this is to use an alligator clip to audio socket adapter such as the one shown in Figure 10-12. These adapters can be found on eBay or Amazon for a few dollars. Having a standard audio socket connected to your micro:bit also means that you can attach a powered loudspeaker to it such as the one shown in Figure 10-13.

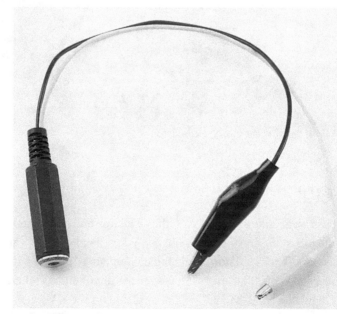

Figure 10-12   *An alligator clip to audio socket adapter.*

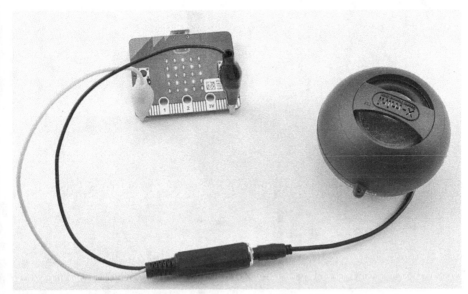

Figure 10-13   *Connecting a powered speaker to a micro:bit.*

Amplifier Modules

Another way to attach a speaker to your micro:bit is to use an amplifier module such as the MonkMakes Speaker for micro:bit (Figure 10-14) or Proto-PIC's amp:bit (Figure 10-15).

The Speaker for micro:bit connects using alligator clips, whereas the amp:bit uses the micro:bit's edge connector, and the micro:bit is plugged directly into the amp:bit. The amp:bit also provides a volume control.

Updating the Alarm Project

You may remember that I promised that the timer project would have an audible alarm added to it. Now that you have several ways of adding sound to your micro:bit, you can add this to the project.

Connect your chosen arrangement of speaker, and then upload the program `ch10_timer_final.py`. Try out the project, and when the timer completes, the alarm will sound with a pleasing tone.

To play the sound, the `music` library must be imported at the top of the program. The other changes are to be found in the `handle_run_state` and `handle_alarm_state` functions.

Figure 10-14   *The MonkMakes Speaker for micro:bit.*

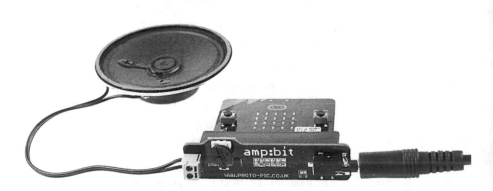

Figure 10-15   *The Proto-PIC amp:bit.*

```
def handle_run_state():
    global state, mins, secs, last_tick_time
    if button_b.was_pressed():
        state = SET
        display_mins(mins)
    time_now = running_time()
    if time_now > last_tick_time + 5000:
        last_tick_time = time_now
        secs -= 5
        if secs < 0:
            secs = 55
            mins -= 1
        display_time(mins, secs)
    if mins == 0 and secs == 0:
        state = ALARM
        music.pitch(440, wait=False)
        display.show(Image.HAPPY)

def handle_alarm_state():
    global state, mins
    if button_b.was_pressed():
        state = SET
        music.stop()
        mins = 1
        display_mins(mins)
```

The music method pitch has a parameter of the pitch (frequency in hertz) of the tone to be generated. By default, this would continue indefinitely, and we would not be able to cancel the tone by pressing a button, so the optional wait parameter is set to False so that our code can detect a press of button B. To stop the music playing when the mode is changed back to SET, the stop method is called. If you decide to use a pin other than pin0 to generate the sound, the play and stop methods must be supplied with an optional pin parameter.

Music

In addition to just playing a tone, the music module has a beautifully designed musical notation that lets you play tunes. A tune is contained in a List of strings, with each string being one note. Only one note can be played at a time. To get the hang of this notation, with your sound hardware attached, open the REPL and try a few commands:

```
import music
music.play(['c4', 'd', 'e', 'f', 'g', 'a', 'b', 'c5'])
```

The first part of the string is the note letter. This can optionally be followed by an octave number.

The `music` module also allows you to specify note duration, sharps and flats, and even a time signature. Rather than rewrite the already excellent documentation on this module, please refer directly to https://microbit-micropython.readthe docs.io/en/latest/music.html.

Speech

Another useful module that you can import into your programs is the `speech` module. Note that although the official documentation for the module (https://microbit-micropython.readthedocs.io/en/latest/tutorials/speech.html) tells you to wire a speaker between `pin0` and `pin1`, if you are using a powered speaker such as the MonkMakes Speaker for micro:bit or a headphone adapter lead, you can connect up normally using `pin0`, as described earlier.

The speech generation sometimes requires some careful listening to hear the message properly, but it is a fun thing to add to a project. Try out the following commands in the REPL:

```
import speech
speech.say("It was the best of times, it was the worst of times.")
```

If you want to vary the pitch or timbre of the voice and even make your micro:bit sing, then take a look at the official documentation for this module.

Neopixels

In addition to simple LEDs such as the red LED used earlier in this chapter and the 25 LEDs that make up a micro:bit's display, there are more advanced LEDs (often called *neopixels*) with built-in chips that allow you to control the color of the LED. These LEDs are designed to be daisy-chained so that you can have a long chain of neopixels, each one connected to the preceding one. A single data pin is used to specify the color each LED is to display by sending a long stream of information.

Fortunately, there is a MicroPython module for using neopixels, and so you don't need to get involved in constructing the data. You can just specify a color for

Figure 10-16   *The Proto-Pic micro:pixel edge.*

each neopixel in the chain. Figure 10-16 shows a very neat neopixel-based display that plugs into the micro:bit's edge connector.

You can buy neopixel displays on a reel of tape that can be cut to the number of neopixels you want, as well as various arrangements of the neopixels, including grids of pixels, even when arranged in a grid on a board. The limitations of the micro:bit's power supply mean that you can only power about eight neopixels from a micro:bit without having to use a separate power supply. You may also have problems powering this project with a 3 V battery pack because unless the batteries are really fresh, there will probably not be enough volts for the neopixels' liking.

When powering neopixels directly from the micro:bit, one data pin is used; the GNDs of the neopixel display and the micro:bit need to be connected, as does the 3 V supply to the neopixels. Neopixels will work at 3 V, but for full brightness, they should be powered at 5 V via a separate power supply.

Flash the program `ch10_neopixels.py` onto your micro:bit. The display will produce a light-chasing effect.

```
from microbit import *
from neopixel import NeoPixel

num_pixels = 8
color = [255, 0, 255] # red, green, blue
```

```
off = [0, 0, 0]

pixels = NeoPixel(pin0, num_pixels)

while True:
    for i in range(0, num_pixels):
        pixels[i] = color
        pixels.show()
        sleep(100)
        pixels[i] = off
        pixels.show()
```

All the code for using the neopixels is contained in the class NeoPixel imported from the built-in module neopixel. When you create an instance of this class, you specify the number of pixels in your display, which for convenience is kept in the global variable num_pixels, so you will need to change this if your neopixel display does not have eight LEDs.

A variable color is used to hold the color to be displayed on the LED. The color is represented as a List of three values. Each value must be between 0 and 255 and represents the amount of red, green, and blue that will make up the color of the LED. So the color [255, 0, 255] is maximum red, no green, and maximum blue, producing a magenta color. The variable off contains the three zeros needed to turn the LED off entirely.

The loop steps over each pixel position in turn, lights the LED in the chosen color, delays, and then turns the LED off again. Note that the change to any pixel pattern is not actually shown until you call pixels.show.

The Edge Connector

In addition to the five alligator lead–friendly connectors on the micro:bit, you will have noticed the much smaller connectors in between the large connectors. These are only accessible by plugging the micro:bit into a socket, such as the one shown on the Proto-Pic micro:pixel edge shown in Figure 10-16.

There are many other types of expansion boards for the micro:bit, including motor controllers, various types of displays, and audio interfaces. You will find suppliers for these and some of their products listed in Appendix A.

In addition to connecting a ready-made expansion board to your micro:bit, if you acquire an adapter, you can gain access to all the pins of the GPIO connector for more advanced projects than are possible with just the pins 0 to 2. You can find a lot more detail about the edge connector at http://tech.microbit.org/hardware/edgeconnector_ds/.

Pinout

Although the edge connector has 25 connections, this does not mean that you get 25 GPIO pins to use. Quite a few of the connections are for GND and 3 V, and many of the connections are also connected to pins that are used to drive the display or for the buttons. Pins that are totally free to use for your own purposes are 0, 1, 2, 13, 14, 15, 16, 19, and 20. Figure 10-17 shows the pinout for the connector.

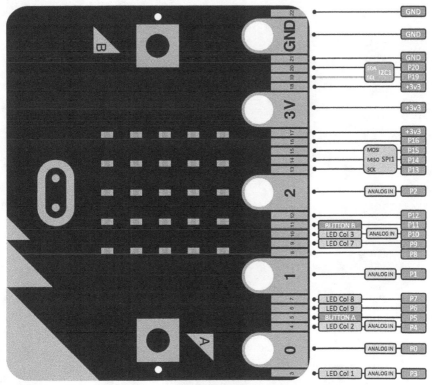

Figure 10-17 *The micro:bit edge connector pinout.*

Pins 0, 1, 2, 3, 4, and 10 can all be used as analog inputs. Pins 13, 14, and 15 can be used as GPIO pins (but not analog inputs) and also can be used for an interface called the *serial programming interface* (SPI), which some chips and modules use.

Similarly, pins 19 and 20 can be used for a more common type of interface called I^2C (pronounced "I squared C"), which also can be used by various types of modules and integrated circuits.

Breakout Boards and Kits

A *solderless breadboard* is a plastic block with holes in the top and metal clips behind the hole to make it easy to connect electronic components such as LEDs, resistors, and photoresistors. To be able to connect the pins from the micro:bit's edge connector to the breadboard, you need an adapter such as the Proto-PIC bread:bit shown in Figure 10-18.

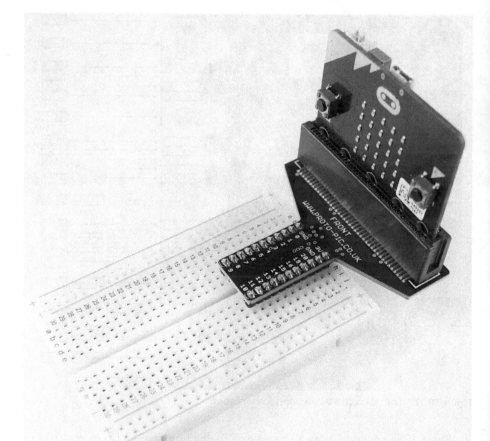

Figure 10-18   *Connecting a micro:bit to a solderless breadboard.*

Taking this a stage further, the Proto-PIC exhi:bit (Figure 10-19) provides a whole platform for connecting to the micro:bit in many ways, including adding alligator clip–friendly connectors for all the pins of the edge connector.

Figure 10-19   *The Proto-PIC exhi:bit.*

Appendix A provides links for other breadboard adapters as well as kits based on an adapter, breadboard, and some components.

Disabling Hardware

As you saw in Figure 10-17, many of the GPIO pins on the edge connector are used for the display and buttons. If you need to use these pins and don't need the display, then you can turn off the display and free up the pins it uses with the command:

```
display.off()
```

This will free up pins 3, 4, 6, 7, 9, and 10.

Summary

In this chapter, you have seen the various ways that you can add some extra electronics to your micro:bit. You will also find lots of project ideas on the Internet.

11

Radio and Communications

The micro:bit includes the hardware for transmitting and receiving data wirelessly. This radio hardware is intended for Bluetooth communication, but Bluetooth is not available in MicroPython (it uses too much memory). However, the good news is that you can still use the radio hardware to send messages wirelessly from one micro:bit to another. This chapter looks at using this radio hardware and also using the micro:bit's USB interface to communicate with a computer.

Basic micro:bit-to-micro:bit Communication

Getting one micro:bit to send a message wirelessly to another is easy, thanks to the built-in `radio` module. You'll need two micro:bits for this, one with the program `ch11_send.py` flashed onto it and one with `ch11_receive.py`. Let's look at the code for sending first.

```
from microbit import *
import radio

x = 0
radio.on()

while True:
    if button_a.was_pressed():
        radio.send(str(x))
        x += 1
```

After the imports, the variable x is initialized to 0. This will be used to count. The next line turns on the radio hardware. The radio can add up to another 10 mA to the current that the micro:bit consumes. The corresponding off method can be used to reduce the power consumption by just turning on the radio when it's needed. The main loop waits for button A to be pressed and then sends the current value of x after first converting it into a string.

Here is the corresponding receiver code from ch11_receive.py:

```
from microbit import *
import radio

radio.on()

while True:
    try:
        message = radio.receive()
        if message:
            display.show(message)
    except:
        radio.off()
        radio.on()
```

The receiver code uses the method receive, assigning the result to message. If there is no message to be received, then receive will return None, but if there is a message, it is shown on the display.

The try/except is there because there is a slight chance (especially in a room full of micro:bits) of transmissions interfering with each other and causing errors. If this happens, the except code restarts the radio by the time-honored method of turning it off and then on again.

Messaging Different Users

In the preceding example, if you had a third micro:bit, it too would receive the message. In this example, you can send messages to a specific micro:bit. To try out this example properly, you really need at least a couple of friends that also have a micro:bit, and each micro:bit will be identified by a number (Figure 11-1).

For example, we could have the same program running on each micro:bit that will be listening for messages and also waiting for a press of button A to cycle through possible recipents and a press of button B to send a message and cause

Figure 11-1 *Sending messages to a particular micro:bit.*

the smile image to be displayed on the recipient. The program for this is in ch11_messenger.py.

Everyone involved in this project needs to select or be allocated a different number. This will be the ID of their micro:bit, and they will need to change the value of the constant MY_ID to match their ID number. It's easiest if the numbers start at zero and run sequentially. The constant NUM_MICROBITS also needs to be set to one more than the highest ID used.

```
from microbit import *
import radio

MY_ID = 0
NUM_MICROBITS = 10
```

The global variable `recipient` indicates the ID to which the message is to be sent. This will be changed by pressing button A.

```
recipient = 0
radio.on()
```

Because it's very easy to mix up a load of micro:bits, as each micro:bit starts, it will briefly display its own ID.

```
display.show(str(MY_ID), delay=200, clear=True)
display.show(str(recipient))
```

The main loop looks for button A to be pressed, and if it is, it adds 1 to `recipient`. If this makes `recipient` greater than or equal to `NUM_MICROBITS`, then `recipient` is set back to zero. If button B is pressed, then the recipient number is converted into a string and then broadcast to all the other micro:bits.

```
while True:
    if button_a.was_pressed():
        recipient += 1
        if recipient >= NUM_MICROBITS:
            recipient = 0
        display.show(str(recipient))
    if button_b.was_pressed():
        radio.send(str(recipient))
        display.show(Image.YES, delay=500, clear=True)
        display.show(str(recipient))
```

The final section of the main loop checks for incoming messages and whether the message matches the ID of the micro:bit, and then it smiles.

```
    message = radio.receive()
    if message and int(message) == MY_ID:
        display.show(Image.SMILE, delay=1000, clear=True)
        display.show(str(recipient))
```

Advanced Radio Settings

Even at its default settings, the micro:bit's radio has quite an impressive range. Testing this outdoors using the test programs `ch11_range_test_tx.py` (transmit) and `ch11_range_test_rx.py` (receive), I found the range to be around 150 yards/meters in line of sight.

You can actually increase this range by changing the power level (the default is 6) and the bit rate (the default is 1 Mbits/s) of the radio using the `radio.config` command:

```
radio.config(power=7, data_rate=radio.RATE_250KBIT)
```

The maximum size of the data sent is by default 32 bytes (or characters), and decreasing this number using the optional parameter length also increases the chances of the message transmission being received successfully, and hence increases the range.

Computer-to-micro:bit Communication

When you use the print function and text appears in the REPL, your micro:bit is sending a message to your computer over the USB lead. You can also send messages in the other direction so that when you type comething in the REPL, it is sent to the micro:bit. In Chapter 6, you saw how this can be useful when debugging a program by running Python commands on the micro:bit. You can also have the Python program running on the micro:bit catch messages typed in the REPL and, say, show them on the display. Program ch11_usb_receive.py shows how you can do this.

```
from microbit import *

uart.init(115200)

while True:
    if uart.any():
        message = uart.readall()
        display.scroll(message)
```

Flash the program onto your micro:bit, open the REPL, and try tying in some text. It should immediately be displayed on the micro:bit.

The micro:bit has an instance variable called uart (universal asynchronos receiver/transmitter) that handles all serial communications, including communicating on the USB link. This has to be initialized to a certain data transfer speed. For the micro:bit's serial communication to the REPL, this needs to be set to 115,200 baud (bits per second).

Inside the main loop, the any method is used to check whether any data have been sent that have not been read yet. If so, the readall method is used to fetch the text, which is then shown on the micro:bit's display.

Remote Control of Your micro:bit Using Python

Although in the preceding section you used the REPL to send messages to the micro:bit, other software, including Python running on your computer, can also send and, for that matter, receive messages.

This ability to send messages back and forth between Python running on your computer and MicroPython running on your micro:bit forms the basis of the Bitio Project created by David Whale. He originally developed bitio to allow a micro:bit to communicate with the game Minecraft so that you can, among other things, use a micro:bit as a controller for the game. You can read all about this in David's excellent book, *Adventures in Minecraft*.

You can use bitio to remote control your micro:bit from a Python program on your computer (e.g., display things, control GPIO pins, etc.) but also do such things as detect button presses and sensor data from the micro:bit and use them in the program on your computer.

Figure 11-2 shows what's going on when you use bitio.

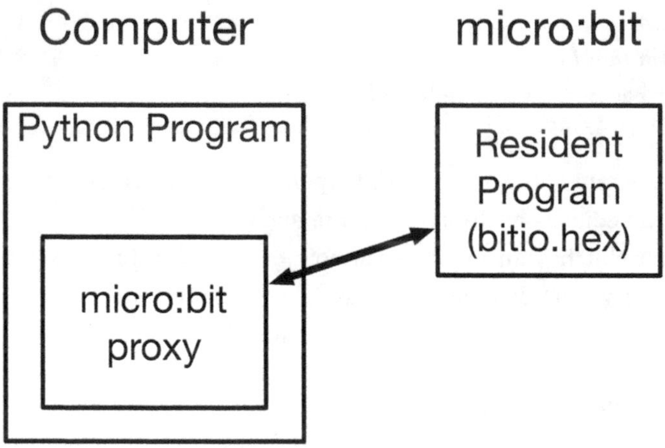

Figure 11-2   *How bitio works.*

To use bitio, you do not need to write aything that actually runs on the micro:bit; you just flash a hex file onto it. This program runs on the micro:bit, providing the link between the micro:bit's hardware and the USB messages between your computer and the micro:bit.

When it comes to using a Python program on your computer, a "proxy" for the micro:bit is used. Thus, when you do things such as this in your computer's Python program, "Hello World" will actually scroll across your micro:bit's display:

```
microbit.display.scroll("Hello World")
```

Python on Your Computer

To use bitio, you first need to get Python running on your computer. The good news is that if you are a Mac or Linux user, your computer already has Python installed. You can test this by starting a terminal session and entering the command python3. This is what I get on my Mac:

```
$ python3
Python 3.6.1 (v3.6.1:69c0db5050, Mar 21 2017, 01:21:04)
[GCC 4.2.1 (Apple Inc. build 5666) (dot 3)] on darwin
Type "help", "copyright", "credits" or "license" for more information.
```

If you are a Windows user, you will need to install Python 3 by following the instructions at www.python.org/downloads/.

Getting Bitio

To install bitio on your computer, you first need to download it from https://github.com/whaleygeek/bitio. On this page, you will see a green button called Clone or Download. Click on this, and select the option Download ZIP (Figure 11-3).

Installing the Resident Program

Unzip the file that you just downloaded, and inside you will find (among other things) the file bitio.hex. Flash this onto your micro:bit.

Using Bitio from the Console

To use bitio either from a Python program or from your computer's command line, your program has to have access to the modules used by bitio. These are all contained in the directory called microbit within a folder called src in the Zip archive that you just downloaded and extracted. So any Python programs that you write for your computer to use bitio either need to be written in that same src

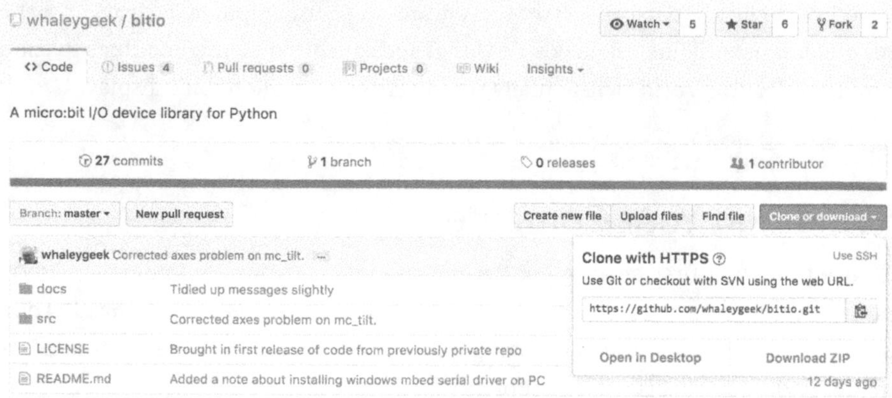

Figure 11-3 *Downloading bitio.*

directory or the whole `microbit` directory needs to be copied into a new directory containing your program.

The quickest way to try out bitio from the command line is to change the directory to the downloaded `src` directory and then run the command `python3`. Do this now, and then import the `microbit` module as shown here:

```
>>> import microbit
No micro:bit has previously been detected
Scanning for serial ports
remove device, then press ENTER
scanning...
found 132 devices
plug in micro:bit, then press ENTER
scanning...
found 133 devices
found 1 new micro:bit
selected:/dev/tty.usbmodem1412
Do you want this micro:bit to be remembered? (Y/N)Y
connecting...
Your micro:bit has been detected
Now running your program
```

The first time that you import the `microbit` module into your computer's Python program, you will be taken through a routine of unplugging your micro:bit and then plugging it back in again so that bitio can detect the port to which your micro:bit is connected. If you accept the option to remember the device, then next time it will just connect automatically.

Once communication has been established, bitio will say `Now running your program`. Try entering the command:

```
>>> microbit.display.scroll("Hello World")
```

You should see the now famous message scroll across the micro:bit's display.

To check communication in the other direction, let's try reading the accelerometer's x, y, and z values.

```
>>> print(microbit.accelerometer.get_values())
(80, 80, -992)
```

You will find documentation and a code example on the GitHub page for bitio mentioned earlier.

Summary

For such a small and low-cost device, the micro:bit's wireless interface is powerful and easy to use. If you have a group of people with micro:bits, you can cook up some truly impressive projects. The micro:bit's USB interface gives you the ability to use the micro:bit's sensors and displays as peripherals to your computer.

This is the final chapter of MicroPython. In Chapter 12, you will find a brief introduction to the JavaScript Blocks editor, the other very common way of programming a micro:bit.

12

JavaScript Blocks Editor

Python is a text-based programming language that is a little more difficult to learn than using a graphical language such as the JavaScript Blocks editor. This is partially because you don't have to worry about getting the punctuation and indenting correct in a block language and also because you don't have to remember much about function and method names because you just pick your blocks of code from lists. In this chapter, you will learn how you can apply much of what you have learned about programming the micro:bit to the JavaScript Blocks editor.

The Editor

The JavaScript Blocks editor works rather like the online version of the Python editor. It all happens on a web page (https://makecode.microbit.org). Open your browser on this page, and after a short delay while a new project is created for you, you will see something like Figure 12-1.

One really nice feature of the Blocks editor is that the image of a micro:bit on the left of the screen is a virtual micro:bit on which you can run your programs before flashing them onto a real micro:bit. You can press its buttons with your mouse; it will display things, and if you used the GPIO pins as digital outputs, it will even highlight them when you write to them. You can also click on GPIO pins to simulate digital and analog inputs.

The middle section of the screen has different categories of blocks: Basic, Input, Music, and so on, where you can find blocks to put onto the right-hand "canvas" area. You can also use the Search box just above the list of block categories if you are not sure where to find the block that you want.

x

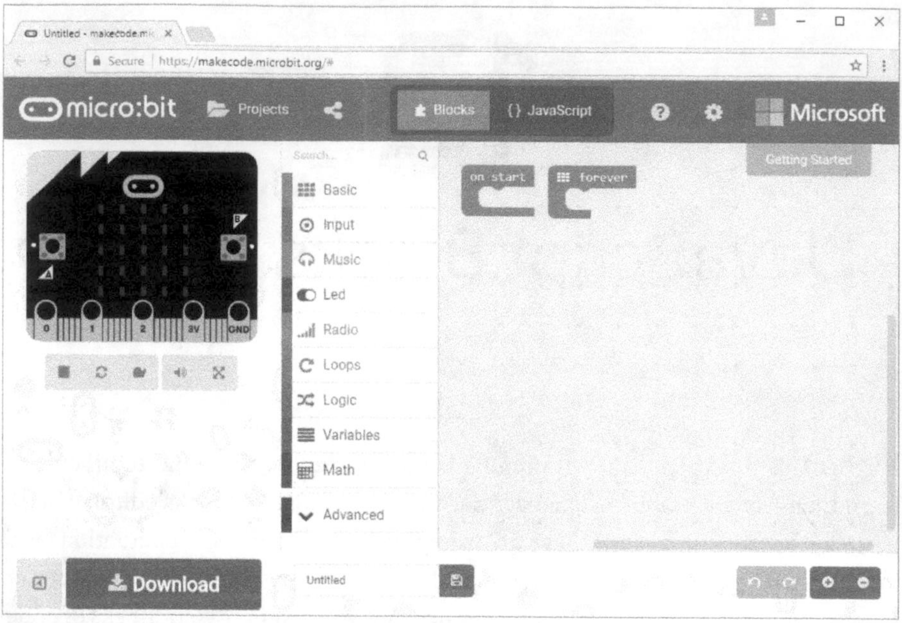

Figure 12-1 *The JavaScripts Blocks editor.*

The Blocks editor actually generates code in the JavaScript programming language, and if you click on the button {} JavaScript, you can see the code that has been generated. Clicking on Blocks takes you back to the blocks view of things, which is much easier to follow.

Getting Started

Let's start where we did with Python and have the display scroll the traditional words "Hello World" and display the heart image repeatedly. When you start a new project, you automatically get on `start` and `forever` blocks. We don't need to do anything special when the project starts, so delete the on `start` block by selecting it and then hitting the DELETE or BACKSPACE key. Next, you are going to add blocks to the `forever` block. Start by adding a `show string` block (in the Basic category). Click on the parameter to `show string` that currently says `"Hello!"` and change it to `"Hello World"`. Your Blocks editor should now look like Figure 12-2.

Notice how the display of your virtual micro:bit is now displaying the message "Hello World". Add some more blocks to the `forever` block: `show icon` and

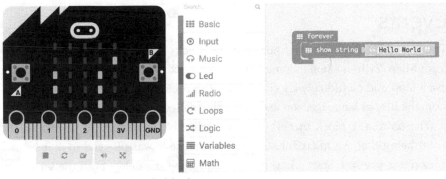

Figure 12-2   *"Hello World" in blocks.*

pause (both of these are in the category Basic). Finally, edit the parameters of the show icon block to an image of your choice and the parameter of pause to be 2000. Your forever block should now look like Figure 12-3, and the virtual micro:bit should be behaving as expected.

If you want to, you can now save your project using the area at the bottom of the window, next to the Download button. Type in the name for your project, and then click on the somewhat anachronistic floppy disk icon to save the project—this will also generate a hex file that you can drag onto your micro:bit in the same way as the online Python editor, as described in Chapter 2.

The Blocks editor will remember any projects that you work on for a particular computer, and you can see them and reopen old projects from the Projects link at the top of the page.

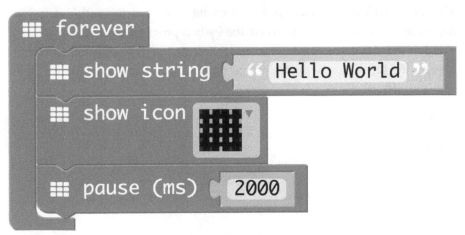

Figure 12-3   *"Hello World" and an image in blocks.*

Events

The JavaScript language that underpins the Blocks editor works rather differently from MicroPython when it comes to things like buttons. Whereas in Python you use a loop and continually check for a button press, in JavaScript and, by extension, the Blocks language, you use something called *events*.

The on start block that we deleted earlier is an example of an event—the event being that the micro:bit started up because it was plugged in or its Reset button was pressed. Start a new project, and this time delete both the on start and forever blocks. Add an on button pressed (another event) block, and then place a show string block inside it, changing the text that the show string block is to display so that it looks like Figure 12-4.

Figure 12-4   *Button events in the Blocks editor.*

Now, when you use your mouse to click on button A in the virtual micro:bit, it will scroll the message. You could also now try this on a real micro:bit.

Variables and Loops

The Blocks language also allows you to create loops and use variables in your programs. For example, let's re-create the Python program ch03_timer_01.py. This was the first step in the timer project, allowing you to set the number of minutes for the timer by repeatedly clicking on button A. Here is the MicroPython code for reference:

```
from microbit import *

mins = 1

while True:
    if button_a.was_pressed():
        mins += 1
        if mins > 10:
```

```
mins = 1
display.scroll(str(mins))
```

The variable `mins` needs to be created in the `on start` block, so from the Variables category, find the `set to` block and place it inside `on start`. Then change the variable name from `item` to `mins` using the Rename option. We also want to display the value of `mins` when it's set, so add a `show number` block to `on start`. To make the `show number` block display the value of the `mins` variable, go to the Variables category, and there you will find the variable `mins` that we just created (Figure 12-5). Drag the variable into the value part of the `set to` block.

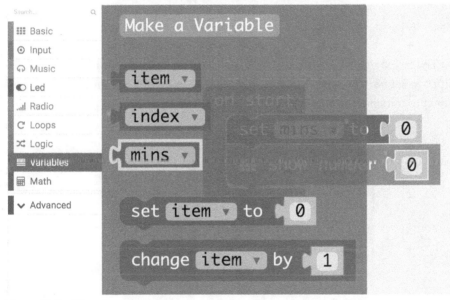

Figure 12-5 *Plugging in a variable.*

We now need an `on button pressed` event, into which you place a `change by` block and change the variable to `mins`. Now your code should look like Figure 12-6.

Next, we need a test to set `mins` to 1 when it gets past 10. This is accomplished by adding an `if then` block (found in the Logic category). For the condition part of the `if then` block, drag in a `condition` block, change the comparison to >, set the right-hand value to 10, and as you did before, add a variable `mins` to the left-hand side of the comparison.

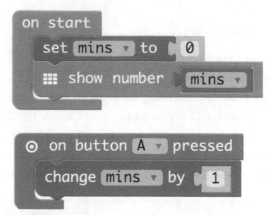

Figure 12-6 *Adding an on button pressed event.*

The last step is to add another set to to the then connector of the if then block to set var to 1 and then copy the show number block from on start. The final program is shown in Figure 12-7.

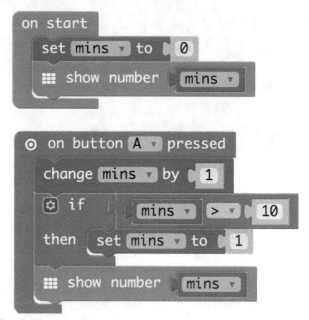

Figure 12-7 *The final version of setting minutes in the timer example.*

Magnetometer Example

You can do pretty much everything with the Blocks editor that you can with Python, and it's a quick and easy language to explore. So we will just look at the magnetometer program that we built in Python to see how this can be built in the Blocks editor.

The Python program in question is ch09_magnetometer.py, and it displays more rows of lit LEDs the closer a magnet is brought to the micro:bit's magnetometer (behind button B). The Blocks version of the program is shown in Figure 12-8.

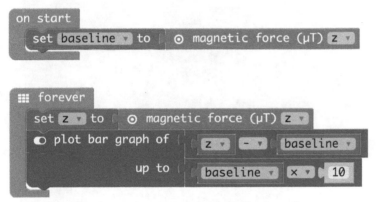

Figure 12-8 *The magnetometer program in Blocks code.*

The Blocks language is very easy to use (including a ready-made way of doing a bargraph-type display), but the payback for this is that it is not possible to define your own functions (although you can do this if you switch to the JavaScript view), also the math functions available are very limited. This could cause a problem for this example, which uses a function to provide inverse-square scaling of the magnetometer readings. However, if you try out the example of Figure 12-8, you will find that the project actually works pretty well without fancy scaling.

Summary

The JavaScript Blocks editor undoutedly provides a really simple way to start doing things with your micro:bit and is a great resource for teaching programming concepts, but it is not without its limitiations.

This is the final chapter, so I will end by pointing you at a few resources that you should find useful:

- http://microbit.org—the starting point for all things micro:bit

- http://microbit-micropython.readthedocs.io/—the official documentation for MicroPython on the micro:bit

- http://tech.microbit.org/hardware/schematic/—for details of how the hardware works, components datasheets, and more

- http://tech.microbit.org/—the home page for the micro:bit developer community

MicroPython Reference

This appendix provides a reference for some of the key features and functions available in MicroPython. Treat it as a reference resource, and try out some of the functions to see how they work. The REPL is your friend. Try things out. For full details on everything in Python, go to https://microbit-micropython .readthedocs.io/en/latest/.

Language Contructs

Table A-1 shows some of the functions you can use with numbers.

Table A-1 *Language Constructs*

Keyword	Description	Example
if	Run a block of code if a condition is true	`if a > 10:` `    print("a is greater than 10")`

if/else		`if a > 10:`
		`    print("a is greater than 10")`
		`else:`
		`    print("a is small")`

if/elif/else		`if a > 10:`
		`    print("a is greater than 10")`
		`elif a > 5:`
		`    print("a is middle-sized")`
		`else:`
		`    print("a is small")`

Keyword	Description	Example

while	Loop while a condition is true	`x = 0`
		`while x < 10:`
		`    x += 1`
		`    print(x)`

for/in	Loop over the items of a List	`for x in range(1, 10):` `    print(x)`

try/except	Run uncertain code and trap errors and exceptions	`try:`
		`    do_something_dodgy()`
		`except:`
		`    print("Something went wrong")`

try/except	Run uncertain code and trap errors and exceptions, with access to the exception	`try:`
		`    do_something_dodgy()`
		`except Exception as e:`
		`    print(str(e))`

try/finally	Mask and ensure cleanup after exceptions.	`try:`
		`    do_something_dodgy()`
		`finally:`
		`    cleanup()`

Comparisons

To test to see whether two values are the same, we use ==. This is called a *comparison operator*. The comparison operators that we can use are shown in Table A-2.

Table A-2 *Comparisons*

Comparison	Description	Example
==	Equals	`total == 11`
!=	Not equals	`total != 11`
>	Greater than	`total > 10`
<	Less than	`total < 3`
>=	Greater than or equal to	`total >= 11`
<=	Less than or equal to	`total <= 2`

Numbers

Table A-3 shows some of the functions you can use with numbers.

Table A-3 *Number Functions*

Function	Description	Example
`abs(x)`	Returns the absolute value (removes any minus sign)	`>>>abs(-12.3)` `12.3`
`bin(x)`	Converts to binary string	`>>> bin(23)` `'0b10111'`
`hex(x)`	Converts to hexadecimal string	`>>> hex(255)` `'0xff'`
`oct(x)`	Converts to octal string	`>>> oct(9)` `'0o11'`
`round(x, n)`	Rounds x to n decimal places	`>>> round(1.111111, 2)` `1.11`
`math.log(x)`	Natural logarithm	`>>> math.log(10)` `2.302585092994046`
`math.pow(x, y)`	Raises x to the power y or use x ** y	`>>> math.pow(2, 8)` `256.0`
`math.sqrt(x)`	Square root	`>>> math.sqrt(16)` `4.0`
`math.sin, cos, tan, asin, acos, atan`	Trigonometry functions, radians	`>>> math.sin(math.pi / 2)` `1.0`

Strings

String constants can be enclosed either with single quotes (most common) or with double quotes. Double quotes are useful if you want to include single quotes in the string like this:

```
s = "It's 3 O'clock"
```

There are other occasions when you may want to include special characters such as end-of-lines or tabs in a string. To do this, you use what is called *escape characters* that begin with a backslash (\). The only ones that you are likely to need are

- \t—tab character
- \n—new line character

Table A-4 lists some of the strings functions that may be useful.

Table A-4 *String Functions*

Function	Description	Example
`s.endswith(str)`	Returns True if the end of the string matches	`>>> 'abcdef'.endswith('def')` `True`
`s.find(str)`	Returns the position of a substring; optional extra arguments of start and end positions to limit the search	`>>> 'abcdef'.find('de')` `3`
`s.format(args)`	Formats a string using template markers using `{}`	`>>> "Its {0} pm".format('12')` `"Its 12 pm"`
`s.isalpha()`	Returns True if all the characters are alphabetic	`>>> '123abc'.isalpha()` `False`
`s.isspace()`	Returns True if the character is a space, tab, or other whitespace character	`>>> ' \t'.isspace()` `True`
`s.lower()`	Converts a string into lowercase	`>>> 'AbCdE'.lower()` `'abcde'`

`s.replace(old, new)`	Replaces all occurrences of "old" with "new"	`>>> 'hello world'.`
		`    replace('world', 'there')`
		`'hello there'`

`s.split()`	Returns a list of all the words in the string separated by spaces; optional parameter to use a different splitting character, the end-of-line character \n being a popular choice	`>>> 'abc def'.split()` `['abc', 'def']`
`s.strip()`	Removes whitespace from both ends of the string	`>>> '    a b    '.strip()` `'a b'`
`s.upper()`	See lower above.	

Lists

Table A-5 summarizes the common List functions.

Table A-5 *List Functions*

Function	Description	Example

del(a[i:j])	Deletes elements from the array from element i to element j−1	```>>> a = ['a', 'b', 'c']```
		```>>> del(a[1:2])```
		```>>> a```
		```['a', 'c']```

a.append(x)	Appends an element to the end of the List	```>>> a = ['a', 'b', 'c']```
		```>>> a.append('d')```
		```>>> a```
		```['a', 'b', 'c', 'd']```

a.count(x)	Counts the occurrences of a particular element	```>>> a = ['a', 'b', 'a']```
		```>>> a.count('a')```
		```2```

a.index(x)	Returns the index position of the first occurrence of x in a; optional parameters for start and end index	```>>> a = ['a', 'b', 'c']```
		```>>> a.index('b')```
		```1```

a.insert(i, x)	Inserts x at position i in the List	```>>> a = ['a', 'c']```
		```>>> a.insert(1, 'b')```
		```>>> a```
		```['a', 'b', 'c']```

a.pop()	Returns the last element of the List and removes it; an optional parameter lets you specify another index position at which to remove	```>>> ['a', 'b', 'c']```
		```>>> a.pop(1)```
		```'b'```
		```>>> a```
		```['a', 'c']```

a.remove(x)	Removes the element specified	```>>> a = ['a', 'b', 'c']```
		```>>> a.remove('c')```
		```>>> a```
		```['a', 'b']```

a.reverse()	Reverses the List	```>>> a = ['a', 'b', 'c']```
		```>>> a.reverse()```
		```>>> a```
		```['c', 'b', 'a']```

a.sort()	Sorts the List; advanced options when sorting Lists of objects	

Dictionaries

Table A-6 lists some common Dictionary functions.

Table A-6 *Dictionary Functions*

Function	Description	Example

`len(d)`	Returns the number of items in the Dictionary	`>>> d = {'a':1, 'b':2}`
		`>>> len(d)`
		`2`

`pop`	Deletes an item from the Dictionary	`>>> d = {'a':1, 'b':2}`
		`>>> pop('a')`
		`1`
		`>>> d`
		`{'b': 2}`

`key in d`	Returns `True` if the Dictionary d contains the key	`>>> d = {'a':1, 'b':2}`
		`>>> 'a' in d`
		`True`

`d.clear()`	Removes all items from the Dictionary	`>>> d = {'a':1, 'b':2}`
		`>>> d.clear()`
		`>>> d`
		`{}`

`get(key, default)`	Returns the value for the key or default if it's not there	`>>> d = {'a':1, 'b':2}`
		`>>> d.get('c', 'c')`
		`'c'`

Type Conversions

Python contains a number of built-in functions for converting things of one type to another. These are listed in Table A-7.

Table A-7 *Type Conversions*

Function	Description	Examples

float(x)	Converts x to a floating-point number	`>>> float('12.34')`
		`12.34`
		`>>> float(12)`
		`12.0`

int(x)	Optional argument to specify the number base	`>>> int(12.34)`
		`12`
		`>>> int('FF', 16)`
		`255`

list(x)	Converts x to a List; also a handy way to get a List of Dictionary keys	`>>> list('abc')`
		`['a', 'b', 'c']`
		`>>> d = {'a':1, 'b':2}`
		`>>> list(d)`
		`['a', 'b']`

B

Hardware

micro:bit Edge Connector Pinout

Figure B-1 shows the pinout for the micro:bit's edge connector. See Chapter 10 for more information on using the edge connector.

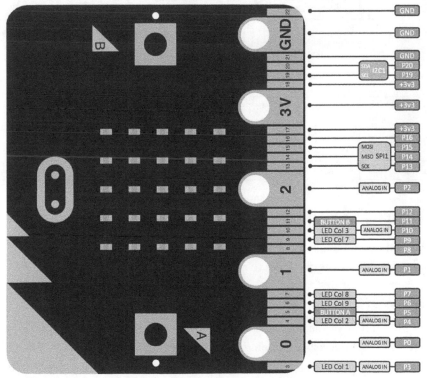

Figure B-1 *The micro:bit edge connector pinout.*

Suppliers and Manufacturers

The micro:bit is available from many suppliers worldwide, so for a micro:bit itself, your best starting point is probably an Internet search. There are also manufacturers and retailers that supply micro:bits and also various add-ons and kits for the boards. Some of the most popular companies are listed in Table B-1.

Table B-1 *Manufacturers and Suppliers of micro:bit-Related Products*

Supplier	Description	URL
Adafruit	A major US retailer of hobby electronic supplies	adafruit.com/
Classroom.com	STEM (science, technology, engineering, and math) resources, primarily for the Chinese market	classroom.com.hk/
CPC	UK-based supplier of components including micro:bit-related products and kits	cpc.farnell.com/
Kitronik	UK-based supplier of STEM resources including its own range of micro:bit products	www.kitronik.co.uk/
Maplin	UK-based bricks-and-mortar as well as online store for hobby electronics	maplin.co.uk
MCM	US-based supplier of components including micro:bit-related products and kits (sister company of CPC)	www.mcmelectronics.com/
MonkMakes	UK-based manufacturer of STEM and maker kits and boards	monkmakes.com
Pimoroni	UK-based supplier of STEM and maker products including its own range of micro:bit products	shop.pimoroni.com/
Proto-PIC	UK-based supplier of STEM and maker products including its own range of micro:bit products	www.proto-pic.co.uk/
SparkFun	A major US retailer of hobby electronic supplies	www.sparkfun.com/

Components

This section contains lists of the various parts and modules used in this book along with some ideas on where to get them.

Hardware

It's a little unfair of me to single out certain products, especially because new products for the micro:bit are being released all the time. What I list in Table B-2 are items that I see as particularly useful. I suggest a thorough Internet search to see the full range of options before you buy anything.

Table B-2 *Hardware for the micro:bit*

Item	Manufacturer/Where to Buy
Alligator (aka crocodile) clip leads	Short leads: Adafruit (1592) Long leads: eBay, most suppliers in Table B-1
Battery box, switched	eBay, most suppliers in Table B-1
Battery box, basic	eBay, most suppliers in Table B-1
Breadboard adapter (bread:bit)	Proto-PIC, CPC
Breadboard adapter (exhi:bit)	Proto-PIC, CPC

Kits

Buying a starter kit is a good way to get started with the basic hardware that you need with electronics on a micro:bit. Table B-3 lists some of the kits I am aware of, although I am sure that other kits also will be available by now.

Table B-3 *micro:bit Electronics Kits*

Part	Manufacturer/Where to Buy
Inventors Kit for the BBC micro:bit	Kitronik/most suppliers
SparkFun Inventors Kit for micro:bit	Sparkfun/Sparkfun
UCreate micro:bit Project Kit	CPC/CPC
MonkMakes Electronics Starter Kit for micro:bit	MonkMakes/Amazon
Starter pack for Proto-PIC exhi:bit	Proto-PIC/Proto-PIC

Basic Components

Getting hold of basic components such as LEDs and resistors can be surprisingly difficult if you visit a large component supplier such as Farnell, Mouser, or Digi-Key. Minimum order values and the sheer range of choices can be confusing.

If you are new to buying electronic components, then it is easier to either buy a kit that includes a basic set of components or use a component supplier such as Adafruit or SparkFun that specialize in the hobby market. eBay is also a great source of low-cost components. Table B-4 lists the few basic components used in this book.

Table B-4 *Basic Electronic Components*

Item	Where to Buy (Product Code)
Red 5-mm LED	Adafruit (299), SparkFun (9590), CPC, MCM
1-kΩ ¼-W resistor	SparkFun (13760), CPC, MCM
Photoresistor	Adafruit (161), SparkFun (9088)
MonkMakes Basic Component Kit (includes the items listed above)	MonkMakes/Amazon

Modules and Expansion Boards

The micro:bit has a lively "ecosystem" with many interesting add-on boards available. Table B-5 lists those used in this book, as well as other interesting boards.

Table B-5 *Modules and Expansion Boards for the micro:bit*

Item	Where to Buy
MonkMakes Speaker for micro:bit	monkmakes.com/mb_speaker
MonkMakes Relay for micro:bit	monkmakes.com/mb_relay
MonkMakes Sensor Board for micro:bit	monkmakes.com/mb_sensor
amp:bit	Proto-PIC
micro:pixel	Proto-PIC
Motor driver board for the BBC micro:bit	Kitronik

Index

Page numbers in italics refer to figures.

Electronics Starter Kit for micro:bit

This kit is designed to complement 'Programming the BBC micro:bit'. It doesn't include a micro:bit, but it does include everything else you need to build some of the projects in this book and get started with your micro:bit electronics adventure.

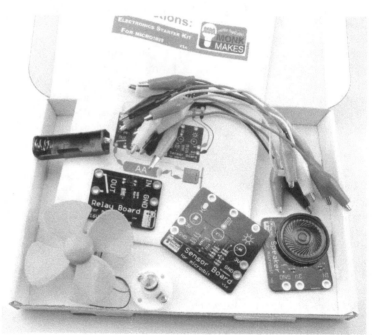

- MonkMakes Speaker for micro:bit
- MonkMakes Relay for micro:bit
- MonkMakes Sensor Board for micro:bit
- Set of 10 short alligator-clip leads
- Small DC motor with fan
- AA battery box
- Flash-light bulb and holder
- Instruction book
- No soldering required

http://monkmakes.com/mb_kit